D1823697

International Cooperation and Arctic Governance

A new exploration of the impacts of Arctic regimes in such vital areas as pollution, biodiversity, indigenous affairs, health and climate change.

The post-Cold War era has seen an upsurge in interest in Arctic affairs. With new international regimes targeting Arctic issues at both the global and regional levels, the Northern areas seem set to play an increasingly prominent role in the domestic and foreign policies of the Arctic states and actors – not least Russia, the USA and the EU.

This volume clearly distinguishes between three key kinds of impact:

- effectiveness, defined as mitigation or removal of specific problems addressed by a regime
- political mobilization, highlighting changes in the pattern of involvement and influence in decision making on Arctic affairs
- region building, understood as contributions by Arctic institutions to denser interactive or discursive connectedness among the inhabitants of the region.

Empirically, the main focus is on three institutions: the Arctic Council, the Barents Euro-Arctic Region and the Council of the Baltic Sea States. This is essential reading for all students with an interest in Arctic affairs and their impact on global society.

Olav Schram Stokke is a Senior Research Fellow at the Fridtjof Nansen Institute, Norway.

Geir Hønneland is Research Director at the Fridtjof Nansen Institute, Norway.

Routledge Advances in International Relations and Global Politics

International Cooperation and Arctic Governance

Regime effectiveness and northern region building

Edited by Olav Schram Stokke and Geir Hønneland

Routledge
Taylor & Francis Group

LONDON AND NEW YORK

First published 2007
by Routledge
2 Park Square, Milton Park, Abingdon, Oxon, OX14 4RN

Simultaneously published in the USA and Canada
by Routledge
270 Madison Avenue, New York, NY 10016

*Routledge is an imprint of the Taylor & Francis Group,
an informa business*

© 2007 Olav Schram Stokke and Geir Hønneland for selection and
editorial matter; individual contributors, their contributions

Typeset in Sabon by
Prepress Projects Ltd, Perth, UK
Printed and bound in Great Britain by
TJI Digital, Padstow, Cornwall

All rights reserved. No part of this book may be reprinted or
reproduced or utilised in any form or by any electronic, mechanical,
or other means, now known or hereafter invented, including
photocopying and recording, or in any information storage or
retrieval system, without permission in writing from the publishers.

British Library Cataloguing in Publication Data
A catalogue record for this book is available from the British Library

Library of Congress Cataloging in Publication Data
International cooperation and Arctic governance: regime effectiveness
and northern regime building / edited by Olav Schram Stokke and
Geir Hønneland.
p. cm. — (Routledge advances in international relations and global
politics: 50)
   Includes bibliographical references and index.
   ISBN 0–415–39934–3 (hardback : alk. paper) 1. Arctic regions—
History. 2. Arctic regions—International status. 3. Arctic regions—
Research—International cooperation. I. Hønneland, Geir. II. Stokke,
Olav Schram, 1961–
G606.I58 2007
320.609 1 1'3--dc22
2006018736

ISBN10: 0–415–39934–3 (hbk)
ISBN10: 0–203–96233–8 (ebk)

ISBN13: 978–0–415–39934–0 (hbk)
ISBN13: 978–0–203–96233–6 (ebk)

Contents

Illustrations

Figures

Tables

Contributors

Alf Håkon Hoel is Associate Professor of Political Science at the University of Tromsø, Norway. His works include studies of international fisheries management, ocean policy development and Arctic affairs. He was co-lead author of the fisheries chapter of the Arctic Climate Impact Assessment. Among his recent publications is *A Sea Change? The Exclusive Economic Zone and Governance Institutions for Living Marine Resources* (Springer, 2005).

Geir Hønneland is Research Director at the Fridtjof Nansen Institute, Norway. He has published widely on Russian environmental politics and international relations in the European North, most recently *Tackling Space: Federal Politics and the Russian North* (University Press of America, 2006), *Law and Politics in Ocean Governance* (Martinus Nijhoff/Brill, 2006), *Health as International Politics* (Ashgate, 2004) and *Russia and the West: Environmental Cooperation and Conflict* (Routledge, 2003).

Kristine Offerdal is writing her doctoral dissertation at the University of Oslo on international oil and gas collaboration in the Arctic, with special emphasis on the United States, the European Union and Russia. She holds a Masters Degree in Administration and Organization Theory from the University of Bergen, with specialization in Russian regional politics. She is currently a Research Fellow at the Fridtjof Nansen Institute, Norway.

Indra Øverland holds a DPhil from the Scott Polar Research Institute at the University of Cambridge, UK. He has worked on a broad range of issues related to energy, aid and indigenous peoples, focusing on the post-Soviet Arctic, the South Caucasus and Central Asia. He has extensive work experience from Norwegian government bodies and is currently Director of the Energy Programme at the Norwegian Institute of International Affairs (NUPI).

Lars Rowe is a Research Fellow at the Fridtjof Nansen Institute, Norway, currently working on a doctoral dissertation on the history of the Pechenga-Nikel Combine on the Kola Peninsula. He has practical experience from international environmental cooperation in the Arctic and has done research on international health initiatives in the Barents and Baltic Sea regions. He is co-author of *Health as International Politics* (Ashgate, 2004).

Peter Johan Schei is Director General of the Fridtjof Nansen Institute, Norway. A former Director General of the Norwegian Directorate of Nature Management, he has wide experience from international environmental cooperation. Among his most recent publications are the edited volumes *Invasive Alien Species: A New Synthesis* (Island Press, 2005) and *China's Protected Areas* (Tsinghua University Press, 2004).

Olav Schram Stokke is a Senior Research Fellow at the Fridtjof Nansen Institute, Norway. His research interests are regime theory, regional cooperation and international management of natural resources and the environment. He has published extensively in these fields, including the edited volumes *Implementing the Climate Regime* (Earthscan, 2005), *Governing High Seas Fisheries* (Oxford University Press, 2001), *Governing the Antarctic* (Cambridge University Press, 1996), and *The Barents Region* (Sage, 1994).

Elana Wilson holds a DPhil in Human Geography from the Scott Polar Research Institute, University of Cambridge, UK. Her interests include Northern economic development and resource use in Canada and Russia, indigenous governance, and the politics of energy and climate change. She is currently working at the Centre for Russian Studies, Norwegian Institute of International Affairs (NUPI).

Acknowledgements

Many of those who have provided comments and advice on this book are acknowledged in the individual chapters. As editors we are especially grateful to Oran Young, who prepared an external review of the volume as a whole, and to Henry P. Huntington, Johnny-Leo Jernsletten, Bernd Rechel, Odd Rogne and David Scrivener for reviewing individual chapters. Like the publisher's anonymous reviewers, they all offered insightful and constructive suggestions on how the drafts could be further advanced.

We would also like to thank Susan Høivik and Maryanne Rygg for excellent editorial services. The project has been funded by the Research Council of Norway and the Fridtjof Nansen Institute.

O. S. S. and G. H.
Oslo, 23 May 2006

Acronyms and Abbreviations

ACAP	Arctic Council Action Plan
ACIA	Arctic Climate Impact Assessment
ACOPS	Advisory Committee on Protection of the Seas
AEPS	Arctic Environmental Protection Strategy
AHDR	Arctic Human Development Report
AMAP	Arctic Monitoring and Assessment Program
AMEC	Arctic Military Environmental Cooperation
ASC	Assessment Steering Committee
BEAC	Barents Euro-Arctic Council
BEAR	Barents Euro-Arctic Region
CAFF	Conservation of Arctic Flora and Fauna
CBD	Convention on Biological Diversity
CBSS	Council of the Baltic Sea States
CLRTAP	Convention on Long-Range Transported Air Pollution
CPAN	Circumpolar Protected Areas Network
CSO	Committee of Senior Officials
DOTS	Directly Observed Treatment with Short-course therapy
EMP	Euro-Mediterranean Partnership
EPPR	Emergency Prevention, Preparedness and Response
EUND	European Union's Northern Dimension
GEF	Global Environment Facility
IASC	International Arctic Science Committee
ICC	Inuit Circumpolar Conference

ICSU	International Council for Scientific Unions
IEA	International Energy Agency
ILO	International Labour Organization
IMO	International Maritime Organization
INRIPP	Institution Building for Northern Russia's Indigenous Peoples Project
IPCC	Intergovernmental Panel on Climate Change
IPS	Indigenous Peoples Secretariat
IWGIA	International Working Group for International Affairs
LOS Convention	United Nations Convention on the Law of the Sea
MARPOL 73/78	International Convention for the Prevention of Pollution from Ships
MFA	Royal Norwegian Ministry of Foreign Affairs
MNEPR	Multilateral Nuclear Environmental Programme in the Russian Federation
NEFCO	Nordic Environmental Finance Corporation
OECD	Organisation for Economic Co-operation and Development
ORPC 1990	Oil Pollution Preparedness, Response and Cooperation
OSPAR Convention	The 1992 Convention for the Protection of the Marine Environment of the North East Atlantic
PAME	Protection of the Arctic Marine Environment
PCB	polychlorinated biphenyls
PINRO	Knipovich Polar Marine Research Institute of Marine Fisheries and Oceanography
POPs	persistent organic pollutants
RAIPON	Russian Association of Indigenous Peoples of the North
Rosgidromet	Russian Federal Service for Hydrometeorology and Environmental Monitoring
SAOs	Senior Arctic Officials

SDWG	Sustainable Development Working Group
SFT	Norwegian Pollution Control Authority
STAKES	National Research and Development Centre for Welfare and Health (Finland)
tcf	trillion cubic feet
UNEP	UN Environment Programme
UNFCCC	UN Framework Convention on Climate Change
WCRP	World Climate Research Program
WHO	World Health Organization
WMO	World Meteorological Organization
WSSD	World Summit on Sustainable Development
WWF	Worldwide Fund for Nature

1 Introduction

Geir Hønneland and Olav Schram Stokke

The post-Cold War era has seen an upsurge of interest in Arctic affairs.[1] This applies to international as well as national politics, and is seen in practical politics as well as in the academic literature. With new international regimes targeting Arctic issues at both the global and regional levels, the northern areas seem set to play an increasingly prominent role in the domestic and foreign policies of the Arctic states and actors – not least Russia, the USA and the EU.

This book is essentially about international governance – or the creation and operation of rules of conduct that define practices, assign roles and guide interaction for dealing with collective problems (Young 1994: 3, 15). For most of the post-war period, institutional means for circumpolar or sub-regional governance across the East–West divide were few and far between, owing to the strategic rivalry between NATO and the Warsaw Pact. This situation changed markedly in the late 1980s, when a series of initiatives were taken for broader cooperation in the Arctic in such vital areas as indigenous issues, communicable disease control, pollution control and biodiversity conservation, climate politics, and environmental concerns in petroleum activities.

Empirically, our main focus here is on three institutions: the Arctic Council, the Barents Euro-Arctic Region and the Council of the Baltic Sea States; the latter is included because salient parts of its Russia-oriented programme activities concern the Arctic oblasts of Murmansk and Arkhangelsk. These relatively recent institutions are functionally broad and address a range of issue areas, often including environmental protection, commerce and industry, health, education and cultural affairs. By contrast, earlier Arctic institutions set up across the East–West divide tended to concentrate on carefully circumscribed issues, such as management of certain shared fish stocks or wildlife populations.[2]

Our focus is on the *impact* of these new Arctic institutions. Their formation and early development have been covered by others.[3] We will pay attention more explicitly to the difference that these institutions might have made. Also, this is a study of impact in quite broad terms. We distinguish between three categories of impact flowing from those institutions: (1) *effectiveness*, defined as mitigation or removal of specific problems addressed by a regime; (2) *political mobilization*, highlighting changes in the pattern of involvement and influence in decision making on Arctic affairs; and (3) *region building*, understood as contributions by Arctic institutions to more dense interactive or discursive connectedness among the inhabitants of the region. These concepts will be further explained in Chapter 2.

The case studies cover some of the many priority areas that the relatively new Arctic institutions have defined for themselves. Especially the sub-regional Barents and Baltic Sea initiatives have placed great emphasis on developing stronger economic ties among the northern territories of the states involved. Programmes under those institutions have aimed at improving physical, financial and administrative infrastructures for commercial and regional business development in areas ranging from fisheries and forestry to maritime transport and telecommunication services. These areas are not covered in this book, or only tangentially.

In terms of theory, the primary challenge is to establish causal substantiation between the operation of Arctic institutions and changes in problem solving, political mobilization or region building. Here it is important to examine whether Arctic institutions succeed in carving out distinctive 'niches' for themselves, as seen in the context of broader international cooperation. While initiatives to Arctic collaboration have often come from the smaller northern states, progress has inevitably depended on support from the 'great powers' in international Arctic politics – Russia, the USA and the EU – whether in political or financial terms.

This introductory chapter provides a brief overview of the Arctic institutions that are in focus in this book. It elaborates on the role of great powers within them, and outlines the chapters that follow.

Cooperative Olympics in the Arctic

The 'Gorbachev initiative', launched by the Soviet leader in a speech in Murmansk in 1987, sparked off a truly hectic period for Arctic policy makers and bureaucrats.[4] The signal that Soviet authorities would welcome more extensive cooperation with Western states on Arctic affairs was quickly heeded, and several scientific communities lost no time in reintroducing an earlier plan for a circumpolar body to foster greater cooperation among Arctic scientific organizations (Roots and Rogne 1987). One prime motivation for this initiative was the desire to obtain physical access for research in the entire circumpolar area. In 1990, these efforts were rewarded by the establishment of the non-governmental International Arctic Science Committee (IASC), an associate of the International Council of Scientific Unions. IASC members are research organizations from all eight Arctic states (Canada, Denmark/Greenland, Finland, Iceland, Norway, Russia, Sweden and the USA) and ten other states. This institution facilitates the development and funding of cooperative projects, in particular those with clear circumpolar relevance.[5]

In 1991, a Finnish initiative to set up a cooperative intergovernmental vehicle for protection of the Arctic environment produced the Arctic Environmental Protection Strategy (AEPS), which includes a string of permanent working groups tasked with various programme activities. The emphasis has been on environmental monitoring, mapping and harmonization of national and international conservation measures, and developing projects for the protection of the marine environment. The four original AEPS activity areas, each coordinated by a working group, have been maintained (now within the broader structure of the Arctic Council): the Arctic Monitoring and Assessment Program (AMAP), Conservation of Arctic Flora and Fauna (CAFF), Emergency Prevention, Preparedness and Response (EPPR), and Protection of the Arctic Marine Environment (PAME).[6] In 1998, the AEPS was incorporated in the Arctic Council, which had been created a few years earlier, following a Canadian initiative and a lengthy period of pushing and shoving.[7] In addition to environmental protection, the Arctic Council addresses social, cultural and economic matters of particular concern to northern communities.[8] Decisions of the Arctic Council are made by bi-annual ministerial meetings, in the format of non-binding declarations that give direction for future work under the Council. The chairmanship rotates among the countries, and the country in the chair is responsible for secretariat functions and driving the cooperation. Day-to-day operations of the Council are taken care of by the countries' Senior Arctic Officials (SAOs), normally polar or Arctic ambassadors. The Arctic Council does not have its own budget or secretariat: work under the Council is dependent upon direct national financial contributions and willingness to act as lead country for projects.

Alongside the development of these circumpolar bodies, two sub-regional institutions have emerged as salient vehicles for cooperation in the European Arctic. A Danish–German initiative gave rise to the Council of the Baltic Sea States (CBSS) in 1992.[9] Member states are the three Baltic states (Estonia, Latvia and Lithuania), the five Nordic countries (Denmark, Finland, Iceland, Norway and Sweden) and Russia, Poland and Germany. The objective of the CBSS is to enhance cooperation and coordination among the states of the region in order to contribute to economic and democratic development in the former Eastern Bloc countries. In addition to these broader goals, specific priority areas are civil security, the fight against organized crime, communicable disease control, environment, labour issues, nuclear safety and transport. Member states are represented by their ministers of foreign affairs. The CBSS takes overall political guidance from the Baltic Sea States Summits, which assemble the heads of government of member states and a member of the European Commission. Chairmanship rotates among the member states. A Committee of Senior Officials (CSO) serves as the body for intergovernmental cooperation among CBSS members between meetings of the Council itself. Decisions within the CBSS are taken by consensus. A permanent international secretariat has been set up in Stockholm.

One year after the founding of the CBSS, in 1993, the Barents Euro-Arctic Region (BEAR) was established on Norwegian initiative. Within a unique two-tiered structure, the Barents Council involves representatives from the governments of Russia, the Nordic states and the EU,[10] whereas its Regional Council involves counties and indigenous peoples in the region.[11] The BEAR facilitates collaboration among the regions of the member states in areas ranging from commerce and industry to culture and education. Its primary objective is to underpin stability and prosperity in the region. More specifically, it aims at reducing military tension, countering environmental threats and addressing the East–West gap in standards of living in the region. Key functional areas are the environment, regional infrastructure, economic cooperation, science and technology, culture, tourism, health care, and the concerns of the indigenous peoples of the region (mainly the Saami and Nenets). Original priority areas concerned infrastructure and environmental and trade collaboration. In terms of project support, however, health care has been by far the most important area for cooperation since the late 1990s.[12] The Barents Council meets at the foreign minister level, with the chairmanship rotating among Finland, Norway, Russia and Sweden. A Committee of Senior Officials (CSO) maintains contact between meetings in the Council. As in the Arctic Council and CBSS, decisions are made by consensus. The four 'core member states' have their own BEAR secretariats, the Norwegian one in Kirkenes being by far the largest. In 2005, Norway launched the idea of turning the

Kirkenes Barents Secretariat into a truly international secretariat for the entire BEAR collaboration.

The idea of a Northern Dimension to the EU was launched by Finland in 1997 and promoted during that country's EU presidency in 1999. Sweden and Denmark pursued the idea during their presidencies in 2001 and 2002. As with the BEAR and CBSS partnerships, the Northern Dimension has a very general ambition – to provide a common framework for the promotion of policy dialogue and concrete collaboration between East and West in Europe. For some time the Northern Dimension remained a rather vague political entity, but it has now come to embrace existing financial EU instruments directed towards the former Eastern Bloc countries: the Tacis, Phare and Interreg programmes. It covers a geographical area stretching from Iceland in the west to northwestern Russia; from the Norwegian, Barents and Kara Seas in the north to the southern coast of the Baltic Sea. Non-EU countries that fall within its scope are Russia, Norway and Iceland. Areas of cooperation under the Northern Dimension include the environment, nuclear safety, energy, infrastructure, trade and business, health and social development.[13]

Associated with the Arctic Council, the CBSS and the BEAR are also a series of meetings among regional and national parliamentarians – the Conference of Parliamentarians of the Arctic Region, the Baltic Sea Parliamentary Conference, and the Barents Sea Parliamentary Conferences, respectively.

Patterns of involvement

These institutions share certain features. All of them have an *international* layer in which the governments of states with Arctic territories seek to develop more extensive collaboration on matters where joint action can be mutually beneficial. Only governments are full members of the Arctic Council, and both the Council of the Baltic Sea States and the Barents Council operate primarily within this layer. In all three institutions, the intergovernmental work is supported by preparatory and implementing meetings of high-level civil servants from relevant ministries: the Arctic Council SAOs and the CSOs of the BEAR and CBSS.

Within the *sub-national* layer, county-level authorities in neighbouring territories separated by national borders seek to coordinate activities and develop joint projects and more unified terms of reference. This layer is particularly salient in BEAR, where representatives from the northernmost counties of Finland, Norway, Russia and Sweden participate in its Regional Council; and a Regional Committee brings together county-level civil servants from all members of the Council. In the CBSS, the sub-national layer of regionality is less prominent, but the Baltic Sea States Sub-Regional Cooperation, established in 1993 and supported by a secretariat, involves more than a hundred sub-national entities. Under this umbrella, they have been recognized as a 'Special Participant' in the CBSS – a status similar to that of observer states.[14] Other sub-national networks with this status are the Union of Baltic Cities and the Baltic Sea Seven Islands (B7 Islands) Cooperation Network. In contrast, the Arctic Council does not have a county-level component, largely because the already existing Northern Forum provided a circumpolar and functionally broad structure for contacts between northern provinces.

The third layer of the institutions examined here is *trans-national* and involves civil society organizations of various kinds – indigenous peoples' organizations in particular. In the Arctic Council, six associations of indigenous organizations, serviced by a secretariat, have status as Permanent Participants, which implies 'active participation and full consultation' in Council work.[15] In BEAR, a representative of the indigenous Saami and Nenets peoples is a full member of the Regional Council; likewise in the Regional Committee. Other civil society organizations are involved in these institutions, but not as prominently. In the Arctic Council, various organizations engaged in scientific, cultural or economic activities participate at working-group or project level, and this is even more true for the Barents and the Baltic Sea regions. In the latter, annual coordination meetings with 'Strategic Partners' have been organized since 2001 by the CSO, to facilitate the involvement of civil society networks and associations in fields such as commerce, industry, human rights and environmental protection.

Arctic institutions and great powers

Three actors in Arctic politics clearly merit the label 'great powers': Russia, the USA and the EU. The specific initiatives for establishment of new Arctic institutions have invariably come from smaller powers – but none of those initiatives could have succeeded without the ultimate support, or at least consent, of the three major actors.

Among the great powers, Russia occupies a particularly central position in Arctic cooperation, and not only because it governs more than half of the Arctic land territories and nearly half of the inhabitants of this region. It was the Soviet policy shift under Gorbachev that made possible enhanced East–West cooperation in the first place, and subsequent initiatives for institution-based region building were all considerably driven by an aspiration to incorporate this country in firm cooperative structures that would also involve its Western neighbours. Most programme activities conducted under these Arctic institutions have been either localized in Russia or motivated largely by economic, social or environmental conditions there. Russia is a full member of all the cooperative institutions dealt with in this book.

Three of the eight member states in the Arctic Council are EU members, as are all the five observer states, but the EU itself does not participate directly in the work of the Council. In contrast, the European Commission is a full member of both the Council of Baltic Sea States and the Barents Council. When the BEAR initiative was launched, none of its states were EU members, but there was recognition that broader European participation would be helpful or even necessary for regional problem solving, given the dimensions of some of the transboundary issues involving northwestern Russia (Castberg *et al.* 1994, Stokke 1994). Actual EU involvement in the BEAR remained low for many years, however (Gjertsen 1997). The 1997 Finnish initiative for developing a Northern Dimension of the EU aimed at changing this situation (Heininen and Langlais 1997) and also sought to strengthen cooperation between the EU and the Arctic Council (Stenlund 2001). While this initiative has yet to produce fresh resources for regional projects, it has aligned several pre-existing financial EU instruments for capacity-building purposes in a geographic area that corresponds largely to the Barents and Baltic Sea regions. For the 2004–6 period, 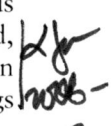 a second Northern Dimension Action Plan has been adopted with an improved structure for implementation that involves annual meetings of, alternately, senior officials and ministers.[16]

The third great power in Arctic politics, the USA, has been somewhat more hesitant in embracing the various initiatives for Arctic cooperation. The establishment of the Arctic Council would have occurred earlier had it not been for US misgivings about the broad mandate assigned to it in early Canadian proposals. Specifically, Washington was determined to ensure that no issues that might touch upon military security would fall within the scope of a new institution.[17] Once the Arctic Council was established, however, the USA began contributing actively to programmes under it, and has assumed leading roles in many Council activities – notably the Arctic Climate Impact Assessment (ACIA) and a range of projects under the CAFF and Sustainable Development umbrellas.[18] Although the USA does not participate in the Baltic Sea region, it is an observer in the Barents Council. Norway's

a high degree of openness regarding BEAR participation
lesire for a solid transatlantic dimension to the Barents
.ration.[19] The USA also engages in the Arctic Military
ntal Cooperation (AMEC), initiated in 1996 and involving
..... uefence ministries of Norway, Russia, and the USA.[20] In 2003, the
United Kingdom joined AMEC.

In this book

Chapter 2 provides a theory platform for the subsequent empirical
discussions. Olav Schram Stokke elaborates a framework focusing on
the regulatory and programmatic responses of Arctic institutions and
their contributions to effectiveness in terms of problem solving, political mobilization and impacts on region building. Among the questions
asked are: What is the causal role of the institutions in any behavioural
change observed? Are there any changes in the pattern of decision-
making involvement within the sector at the regional or the national
level? Have the institutions contributed to increasing the flow of people and ideas within the Arctic region, or promoted the image of the
Arctic as a distinctive political unit? The chapter also provides a background for understanding institutional interplay and niche making in
international Arctic politics.

Case studies from some of the most important fields covered by the
various Arctic institutions follow in Chapters 3 to 7. Elana Wilson and
Indra Øverland start out with an examination of indigenous issues in
Chapter 3. Indigenous problems have been addressed by the Arctic
Council and the BEAR in different ways. Whereas indigenous organizations have used their relatively strong position in the Arctic Council
to influence the behaviour of states and other international institutions,
the primary focus in the BEAR has been aid and development projects
among the Russian Saami. The chapter maps these developments and
discusses the niches made by the Arctic Council and the BEAR in policy
making on indigenous affairs.

Health issues came to the fore in international Arctic collaboration
in the late 1990s when alarming figures emerged about the spread of
new and re-emerging communicable diseases, especially HIV/AIDS
and tuberculosis, from Russia's northern regions. In Chapter 4, Lars
Rowe and Geir Hønneland discuss the main experiences of the Barents
Health Programme (BEAR) and the Task Force on Communicable Disease Control in the Baltic Sea Region (CBSS). Both initiatives have focused on improving the capabilities of post-Soviet states for halting the
spread of communicable diseases, particularly through introducing the
World Health Organization's (WHO) regime for tuberculosis control.

Rowe and Hønneland pay special attention to how these efforts have been received in the post-Soviet states.

Pollution was the classical environmental problem in focus during the first years of the post-Cold War period. Especially the Russian sector of the Arctic was hard hit by the effects of industrial pollution, and this situation evoked political and financial responses from many international institutions. In Chapter 5, Olav Schram Stokke, Geir Hønneland and Peter Johan Schei look at how these efforts have progressed, and how Arctic institutions have handled issues of pollution and conservation more generally.

Climate change is another topic that has ranked high on the political agenda of Arctic institutions, especially the Arctic Council, in recent years. Changes in the Arctic climate have global ramifications, and the effects of climate change are particularly severe in the Arctic. Is it getting warmer, wetter, wilder in the Arctic? asks Alf Håkon Hoel in Chapter 6. He focuses on how the ACIA was organized under the auspices of the Arctic Council on the joint initiative of various programmes within the Council and other Arctic institutions, such as the IASC. The chapter reports how the ACIA was carried out from 2000 to 2004 and discusses how this relates to the global climate regime and other scientific processes in the Arctic.

The extraction and transport of oil and gas constitute a new potential source of environmental degradation in the Arctic as technological advances and a warmer climate make the hydrocarbon resources of the region more accessible. Both the USA and the EU are increasingly keen on the gas expected to flow from the Russian part of the Barents Sea in the near future. Are Arctic institutions prepared to meet this challenge? This is the question Kristine Offerdal asks in Chapter 7. She reviews relevant documents produced and forthcoming by the Arctic Council and its programmes, like the AMAP Oil and Gas assessment. Again, at the heart of the discussion is the special niche for Arctic institutions in more global processes.

In Chapter 8, Olav Schram Stokke summarizes the findings of those five case studies with respect to regime impacts on effectiveness, political mobilization, and Arctic regionality. Particular attention is paid to the interaction between Arctic institutions and other levels of governance.

Notes

1 We would like to thank David Scrivener, Oran Young and our fellow contributors to this book for very helpful comments. We apply the spatial boundary of the Arctic defined in the *Arctic Human Development Report*

(AHDR 2004: 17–18), which slightly adapts that used by the Arctic Monitoring and Assessment Programme (AMAP 1997: 7). This encompasses all of Alaska; Canada north of 60° and northern Quebec and Labrador; all of Greenland, the Faroe Islands and Iceland; the northernmost counties of Norway, Sweden and Finland; and in Russia, the Murmansk oblast and parts of Arkhangelsk oblast, the Nenets, Yamalo-Nenets, Taimyr and Chukotka autonomous areas, and parts of the Komi Republic, Krasnoyarsky Kray and the Sakha Republic.

2 On Norwegian–Russian fisheries cooperation on Barents Sea fish stocks, see Stokke *et al.* (1999) and Hønneland (2000, 2004); on the Polar Bear Convention, see Fikkan *et al.* (1993); on the North Pacific Sealing Convention, see Mirovitskaya *et al.* (1993).

3 See, e.g., Stokke and Tunander (1994), Young (1998), Tennberg (2000) and Keskitalo (2004), among others.

4 For an overview of the Gorbachev initiative and early responses to it, see Scrivener (1988).

5 Details on membership, procedures, funding mechanisms and project portfolio are available at www.iasc.no. The significance of the IASC process for the emergence of the concept of 'the Arctic eight' states is emphasized in Keskitalo (2004: 45–7).

6 Declaration on an Arctic Environmental Protection Strategy (1991). On the formation of this institution, see Scrivener (1996) and Young (1996). For details on current projects, see www.arctic-council.org.

7 Declaration on the Establishment of the Arctic Council (1996). On the process of establishing this institution, which began around 1990, see Young (1996).

8 In addition to the activity clusters defined under AEPS, Arctic Council projects are coordinated by its Sustainable Development Working Group (SDWG) and the Arctic Council Action Plan to Eliminate Pollution of the Arctic (ACAP); see www.arctic-council.org.

9 Declaration from the Conference of Foreign Ministers of the Baltic Sea States (1992). Information about history, structure and programmatic foci is available at www.cbss.st.

10 Declaration on Cooperation in the Barents Euro-Arctic Region (1993). Analyses of formation and early operation of this institution are provided in Stokke and Tunander (1994).

11 At the regional level, the BEAR initially included the three northernmost counties of Norway, Norrbotten in Sweden, Lapland in Finland, Murmansk and Arkhangelsk in Russia. The BEAR's geographical scope has subsequently been extended to include the counties of Västerbotten (Sweden), Oulu and Kainuu (Finland) and the Republics of Karelia and Komi in Russia. Nenets Autonomous Okrug, located within Arkhangelsk Oblast in Russia, became a member in its own right in 1997.

12 Information on institutional structure, membership and programme activities is available at www.beac.st.

13 An overview of the EU Northern Dimension is found at http://europa. eu.int/comm/external_relations/north_dim/.

14 The Baltic Sea Sub-Regional Cooperation is one of six so-called 'Special Participants' that since 1999 may participate on an *ad hoc* basis in work

under the CBSS and its working groups; see Principles and Guidelines for Third Party Participation, available at www.cbss.st.

15 Declaration on the Establishment of the Arctic Council (1996), Art. 2. The three original Permanent Participants were the Inuit Circumpolar Conference, Saami Council and the Russian Association of Indigenous Peoples of the North; subsequently, the Aleut International Association, the Arctic Athabaskan Council and the Gwich'in Council International have obtained this status.

16 An overview of the Northern Dimension of the EU, including the Second Action Plan, is available at www.europa.eu.int/comm/external_relations/north_dim.

17 Declaration on the Establishment of the Arctic Council (1996), Art. 1, states (in a footnote) that the Council 'should not deal with matters related to military security'.

18 See www.arctic-council.org.

19 See Castberg *et al.* (1994). Membership in the Barents Council is open to any state wishing to take an active part, but chairmanship will rotate between the four states governing the counties involved in the cooperation; see Terms of Reference for the Council of the Barents Euro-Arctic Council, Arts 2 and 6.

20 Declaration on Arctic Military Environmental Cooperation, Bergen, 26 September 1996.

References

AHDR (2004) *Arctic Human Development Report*. Akureyri: Stefanssson Arctic Institute, under the auspices of the Icelandic Chairmanship of the Arctic Council, 2002–4.

Castberg, Rune, Olav Schram Stokke and Willy Østreng (1994) 'The dynamics of the Barents Region' in O. S. Stokke and O. Tunander (eds.), *The Barents Region: Cooperation in Arctic Europe*, London: SAGE, pp. 71–84.

Declaration from the Conference of Foreign Ministers of the Baltic Sea States (Copenhagen, 6 March 1992). www.cbss.st.

Declaration on an Arctic Environmental Protection Strategy (Rovaniemi, 14 June 1991). www.arctic-council.org.

Declaration on Cooperation in the Barents Euro-Arctic Region, Conference of Foreign Ministers (Kirkenes, 11 January 1993). www.beac.st.

Declaration on the Establishment of the Arctic Council (Ottawa, 19 September 1996). www.arctic-council.org.

Fikkan, Anne, Gail Osherenko and Alexander Arikainen (1993) 'Polar bears: the importance of simplicity' in O. R. Young and G. Osherenko (eds.), *Polar Politics: Creating International Environmental Regimes*, Ithaca, NY: Cornell University Press, pp. 96–151.

Gjertsen, Arild (1997) *Norge og Russland – fragmenterte pragmatikere?* Unpublished dissertation, Department of Political Science, University of Oslo.

Heininen, Lassi and Richard Langley (eds.) (1997) *Europe's Northern Dimension: The BEAR Meets the South*, Rovaniemi: University of Lapland.

Geir (2000) *Coercive and Discursive Compliance Mechanisms in* *agement of Natural Resources: A Case Study from the Barents Sea* , Dordrecht: Kluwer Academic.

Geir (2004) *Russian Fisheries Management: The Precautionary Ap-* *ᴨ Theory and Practice*, Leiden: Martinus Nijhoff.

Keskitalo, E. C. H. (2004) *Negotiating the Arctic: The Construction of an International Region*, New York: Routledge.

Mirovitskaya, Natalia S., Margaret Clark and Ronald G. Purver (1993) 'North Pacific fur seals: regime formation as a means of resolving conflicts' in O. R. Young and G. Osherenko (eds.), *Polar Politics: Creating International Environmental Regimes*, Ithaca, NY: Cornell University Press, pp. 22–55.

Roots, E.F. and O. Rogne (1987) 'The Need for Feasibility and Possible Role of an International Arctic Science Committee', memo, Oslo: Norwegian Polar Institute.

Scrivener, David (1988) 'Gorbachev's Murmansk speech: Soviet initiative and Western responses', *Security Policy Library*, No. 1, Oslo: The Norwegian Atlantic Committee.

Scrivener, David (1996) 'Environmental cooperation in the Arctic: from strategy to council', *Security Policy Library*, No. 1, Oslo: The Norwegian Atlantic Committee.

Scrivener, David (1999) 'Arctic environmental cooperation in transition', *Polar Record*, 35: 51–8.

Stenlund, Peter (2001) 'The Arctic window of the Northern Dimension' (www. arctic-council.org/en/main/infopage/31/), accessed 29 March 2005.

Stokke, Olav Schram (1994) 'Environmental cooperation as a driving force in the Barents Region' in O. S. Stokke and O. Tunander (eds.), *The Barents Region: Cooperation in Arctic Europe*, London: SAGE, pp. 145–59.

Stokke, Olav Schram and Ola Tunander (eds.) (1994) *The Barents Region: Cooperation in Arctic Europe*, London: SAGE.

Stokke, Olav Schram, Lee G. Anderson and Natalia Mirovitskaya (1999) 'The Barents Sea fisheries' in O. R. Young (ed.), *The Effectiveness of International Environmental Regimes: Causal Connections and Behavioral Mechanisms*, Cambridge, MA: MIT Press, pp. 91–154.

Tennberg, Monica (2000) *Arctic Environmental Co-operation: A Study in Governmentality*, Aldershot and Burlington, VT: Ashgate.

Young, Oran R. (1994) *International Governance: Protecting the Environment in a Stateless Society*, Ithaca, NY: Cornell University Press.

Young, Oran R. (1996) *The Arctic Council: Marking a New Era in International Relations*, New York: The Twentieth Century Fund.

Young, Oran R. (1998) *Creating Regimes: Arctic Accords and International Governance*. Ithaca, NY, and London: Cornell University Press.

2 Examining the consequences of Arctic institutions

Olav Schram Stokke

Introduction

How should we go about evaluating whether Arctic institutions matter significantly to governments and others in their efforts to cope with regional challenges? In Chapter 1, we identified three types of impact that are especially relevant to the institutions examined here: (1) effectiveness, defined as mitigation or removal of specific problems; (2) political mobilization, highlighting changes in participation and influence in decision making on Arctic affairs; and (3) region building, understood as contributions by Arctic institutions to denser functional or discursive connectedness among the inhabitants of the region. 'Functional connectedness' refers to flows of commodities, people and ideas, whereas 'discursive regionality' is the extent to which the region is thought and spoken of as a distinctive unit.[1] The actors involved in the creation and operation of Arctic institutions have differed in the emphasis they place on each of those categories of impact, but all three figure prominently in the rhetoric surrounding cooperative initiatives.

In developing an analytical framework for the case studies that follow, this chapter proceeds in four steps. First, I pinpoint some challenges associated with substantiating causal connections between Arctic regimes and any changes in problem solving, political participation or regionality. There are various ways of making such substantiation, but all involve serious consideration of rival explanations for such changes. Then I elaborate on the concept of regime effectiveness, notably the objects one may focus on when assessing effectiveness, and the need for explicit evaluative yardsticks. Because many of the problems addressed by Arctic institutions are also targeted by other international bodies, some comments are made about institutional interplay. The third step is to explain how political mobilization on Arctic affairs may be influenced by certain structural features of Arctic institutions. Finally, I discuss relationships between the functional and discursive dimensions of regionality and relate them to activities under Arctic institutions.

Making a difference

Practitioners and scholars pay attention to international institutions because they are believed to contribute to the attainment of desired goals. Much of the early work on international institutions focused on regime formation and maintenance (Krasner 1983). Indirectly, these studies also dealt with regime consequences, by highlighting how international institutions enable states to coordinate their actions in mutually beneficial ways (e.g. Keohane 1984). Not until around 1990, however, were systematic and empirically based inquiries into the actual impacts of international regimes launched; and it is only recently that several more ambitious investigations into the consequences that flow from regimes have been reported.[2]

It follows from the causal nature of a claim about regime impacts – i.e. that an institution matters significantly in solving specific problems, empowering new actors, or strengthening interactive or discursive regionality – that simply observing change along those dimensions will not suffice. Substantiating a causal effect requires examination of other causal processes that might be at work alongside the institution and that could suggest alternative explanations. In an effectiveness assessment, for instance, the problem addressed – whether defined in terms of the environmental, economic, or societal state of affairs – will normally be affected by a number of factors other than social institutions, like changes in demand and supply, technological innovations, and societal beliefs. Moreover, with social institutions, the institution under study is usually only one among several.

One way to assess causal impacts in such complex situations is to trace certain processes that mediate between the institution and the consequence under scrutiny. This approach is often taken in studies involving a small number of cases (Yin 1989). Here evidence for a particular causal account is built by bringing out the fine details of how an outcome came about through a sequence of events, each of which has a causal history less complex than the ultimate outcome and hence is easier to control (George and McKeown 1985). To be persuasive, such tracing must relate the institution to causal mechanisms, i.e. certain constellations of agency and context generally believed to trigger behavioural adaptation (Stokke 2004). In fact, most of the large-scale trans-national projects on regime effectiveness conducted since the early 1990s have been structured as intensive, loosely comparative case studies that embrace such a 'mechanism approach' to the formulation and substantiation of causal claims.

The sets of mechanisms proposed in regime effectiveness studies have varied considerably in specificity and comprehensiveness, but at the high end of generality they can be subsumed under three categories. First, regimes may affect behaviour by cognitive means, i.e. by influencing actors' awareness about certain problems and how they are affected by them, or their knowledge about ways in which problems can be mitigated. Second, regimes may affect perceptions about what is right and proper conduct within an issue area: that is, by making regime norms more compelling. There can be various reasons for such compellingness. Franck (1990: 16) defines legitimacy in the international domain as 'a property of a rule or rule-making institution which itself exerts a pull towards compliance on those addressed normatively because those addressed believe that the rule or institution has come into being and operates in accordance with generally accepted principles of right process'. He goes on to extract from the legal literature on international legitimacy certain 'building blocks of due process', including various types of procedural validation which in the context of environmental issues would highlight matters like transparency and involvement of stakeholders or relevant impartial expertise (Bodansky 1999). A third general way in which regimes may affect behaviour is by altering the utility that actors assign to behavioural options within an issue area, for instance by providing incentives for rule adherence or by adding costs to non-compliance.[3]

The stylized model of behavioural drivers implied by this trichotomy is simple, but the cognitive and normative components sensitize the analysis to processes which are frequently downplayed in the study of international relations: those associated with legitimacy and changes in beliefs and perceived interests. While they will not be equally relevant in each given case, such a broader range of mechanisms can be useful for substantiating the impacts of international regimes on specific problem solving, mobilization or changes in discursive or interactive regionality.

Effectiveness and interplay

The effectiveness of international regimes is usually understood as the extent to which an institution helps in solving or ameliorating the specific problem it was set up to address.[4] Such problem solving is a matter of degree; it is not unusual that changes occur over time, as a result of a gradual strengthening of the regime in question, whether because of more elaborate or more restrictive rules, or because greater resources are made available for regional programmes.

Views differ as to what should be in focus when measuring degrees of problem solving. David Easton's distinctions between outputs (decisions), outcomes (behaviour) and impacts (on policy objectives) have been used to characterize the various objects that may be focused on (Underdal 1992). Early contributions had a distinct output focus: environmental institutions, for instance, were evaluated by the extent of environmental protection activities conducted within them (Kay and Jacobson eds. 1983). Later on, the outputs in focus were broadened to include also state-level activities taken in response to international regimes. By encouraging empirical investigation of relevant activities conducted at international (regime) or national levels, the output approach to regime effectiveness is already far removed from the 'old institutionalism' which concentrated on the formal structure of the bodies involved (Thelen and Steinmo 1992). It would be difficult to find effectiveness studies today that try to do without analysis of institutional outputs.

That said, many scholars highlight the need to go beyond institutional and governmental outputs and examine critically the causal relevance of those outputs to furthering the goals pursued by the regime. This requires expanding the analysis to include the behavioural outcomes that flow from regime-induced rules or programmes. The range of actors whose behaviour is examined will depend on the purpose of the regime and the particular challenges faced in the activity system that a regime seeks to influence. Always included is the behaviour of target groups – those actors, frequently industrial groups, who engage in activities that generate the problems addressed by the regime. As shown by Stokke (1998), for instance, various efforts within Arctic institutions to deal with the threat of nuclear contamination stemming from Soviet-era dumping of nuclear reactors have focused on engaging parts of Russia's military complex in monitoring and preventive activities.[5]

Important as it is to elicit behavioural change in desired directions, in the end regimes are intended to solve, or at least ameliorate, certain socio-economic, political or environmental problems that regional players are struggling with. To evaluate whether behavioural impacts contribute to such improvement, we need explicit yardsticks of effectiveness. Obviously, such yardsticks must reflect the stated objectives of Arctic institutions, or of regime participants, but there are other concerns as well. Notably, the yardstick should mirror the extent to which the behaviour that generates regional problems is within the reach of Arctic institutions or is regulated under other institutions. The interconnectedness of many Arctic challenges (for instance in the environmental area) with activities that occur in other parts of the globe suggests that the effectiveness of Arctic institutions will be contingent on their interplay with other, usually broader, institutions that address the same or closely related issues.

Systematic attention to such institutional interplay, in which the contents or operation of one institution are significantly affected by another, is relatively recent.[6] Whereas the regime literature in general has been criticized for painting a rosy picture of world affairs, with insufficient attention to turf struggles and conflictual interests (Mearsheimer 1995), early studies of institutional interplay were often rather negatively framed. Typical points of departure were instances of normative discord, duplication of work and institutional competition; hence the frequent appearance in the literature of negatively charged terms like 'treaty congestion' and problems associated with increased 'regime density'.[7] One reason for this negative framing lay in certain widely publicized cases of inter-regime tension along the trade–environment border (Runge 1994; Schoenbaum 1997). In contrast, an important finding from recent empirical studies of regime effectiveness, conducted in a wide range of issue areas, is that cross-regime inefficiency or discord are not predominant aspects of interplay.[8]

The approach to institutional interplay taken in this book employs the notion of institutional niches. In ecology, a 'niche' denotes the position of a species or population in an ecosystem, notably that segment of a resource domain where it out-competes other local populations. Used as a metaphor in organizational analysis, the niche concept highlights the relationship between institutional features and ability to extract the resources necessary for organizational survival (Hannan and Freeman 1977). According to the principle of competitive exclusion, no two species can occupy the same niche for a long time, because competition between species or populations will force the weaker party either to adapt, by carving out another niche, or to abandon the ecosystem. It would be imprudent to assume that the competitive exclusion principle operates as forcefully in the realm of social institutions as in natural ecosystems. Nevertheless, it is usually expected of new institutions, like those examined in this book, that they add unique value compared to existing problem-solving efforts. Without such added value, institutions will be vulnerable to charges of wastefulness or redundancy which may ultimately undermine support for their maintenance.

Applied to our cases, a niche approach to interplay inquires whether the Arctic institutions concentrate on aspects of problem solving that are not adequately dealt with by other institutions or that Arctic bodies are particularly well equipped or positioned to address. Here, the three mechanisms of influence can be useful in structuring the analysis. As shown in the case studies that follow, the Arctic institutions have all carved out a *cognitive* niche for themselves, in that generation of knowledge not provided elsewhere has been an important task and meant to form a basis for taking action in areas such as radioactive pollution, the spread of hazardous contaminants, or the state of public health. Only rarely have these institutions sought to occupy a *normative* niche, which would imply active moves to strengthen the contents or application of existing international rules or establish new ones. With respect to the third, utilitarian mechanism, the following chapters provide several instances where Arctic institutions serve to enhance the *capacity* of states and other actors to behave in ways that can further problem solving.[9]

To summarize, an assessment of institutional effectiveness will start out from regulatory and programmatic activities and delve as deeply as possible into their impacts on the behaviour of actors relevant to the problem that the institution was set up to address, and the state of that problem. The parallel emergence of several international institutions attending to a broad range of Arctic challenges, and the fact that many Arctic problems originate in other parts of the world, suggest that institutional interplay will impinge upon the effectiveness of Arctic institutions. Examining such interplay in terms of institutional niches means inquiring whether Arctic institutions focus on aspects of problem solving where they have particular advantages compared to other international bodies.

Political mobilization

For purposes of empirical analysis, the narrow, issue-specific lens of regime effectiveness analysis is a strength as well as a liability. Other things being equal, a sharp focus on specific problem solving permits better causal substantiation of regime impacts. Increasingly, however, students of international regimes have begun exploring a broader set of consequences, including whether regimes influence patterns of decision making between and within states, regional cohesion, or the conditions for peaceful change (Underdal and Young 2004). In this book about Arctic institutions and regional cooperation, the question of participation in decision making is salient, notably whether regional institutions affect bureaucratic or societal involvement.

The structural component of Arctic institutions involves how differentiate among current and potential participants with respect t substantive rights and duties or procedural roles in decision making.[10] In Chapter 1 we pinpointed the vertical aspect of this: how the new institutions serve to connect the various layers of Arctic regionality – the international, trans-national, and sub-national. It is one of the distinctive features of Arctic institutions that they involve not only representatives of national governments but also provincial governments, organizations of indigenous peoples where appropriate, and other civil society groups. This participatory expansion originated with the Finnish initiative to establish the Arctic Environmental Protection Strategy. Prior to this, the Saami Council had for years – and in vain – sought formal involvement in the various regional bodies of Nordic cooperation. In 1989, three indigenous peoples' organizations, among them the Saami Council, had at Canadian insistence been invited to participate as observers in the set of meetings that created the AEPS. This circumpolar process created a path that would have been difficult to deviate from when Norway subsequently launched the BEAR initiative. Indeed, recognition of the legitimacy of indigenous participation has arguably been brought one step further in the BEAR's Regional Council, since no differentiation is made between the indigenous peoples' representative and the county-level leadership assembled in this body, designated as the 'engine' of the Barents Region.[11] Similar comments apply to the participation of the European Commission in the Barents Council and in the Council of the Baltic Sea States. In the early 1990s, all the frontier regions in Europe straddling the old East–West border had explicitly aimed at arousing the interest of the European Union – but only those two succeeded in eliciting a formal EU response (Veggeland 1994). The Commission does not, however, have a formal role in the work of the Arctic Council.

The horizontal dimension of how Arctic institutions structure their decision making is also of interest and stems from the broad functional scope of these new institutions. Covering a wide range of issue areas, especially the sub-regional institutions are well placed to allow various issues to be integrated when considering options for regional projects. In the formative stages of the BEAR, for instance, a subtle linkage was apparently drawn between the realization of broader economic cooperation in the region (desired by Russia), and progress in the environmental area (a Nordic priority) (Stokke 1994).[12] Foreign ministry representatives are the leading governmental players in all the institutions examined here, but field ministries are involved in delegations and at working-group level. A review of the Arctic Council structure concluded that the active and strong involvement of a range

be decisive for its success (Haavisto 2001: 36). The ... ie Barents Council and the Council of the Baltic Sea ... le for separate meetings of field ministers.

Another type of broader consequences important for evaluating Arctic institutions concerns the extent to which they contribute to constructing the Arctic as a political region. The term 'region', when used about political entities, is closely linked to the study of integration – a growth industry in the 1960s. Trying to condense past efforts to define political regions, Thompson (1973) first identified 21 much-cited and rather different meanings. He then restricted the term to clusters of at least two actors, finally narrowing in on geographic proximity, interaction and recognition. It would violate everyday language to term a cluster of territories 'regional' unless they were linked by either land or water. Similarly, no region has been claimed that does not involve considerable interaction within one or more spheres, whether conflictual or cooperative. This does not imply, of course, that the territories in question must be self-sufficient in economic, political or other terms; or that flows of communication and trade beyond regional boundaries are insignificant. What it does mean is that individuals, groups and state representatives view problems and opportunities in the area through a regional prism, and that outsiders too recognize the region as standing out from its surroundings.

In essence, therefore, regionality is the interactive and discursive distinctiveness of more or less clearly defined geographic areas. Moreover, regionality is not simply a condition to be assessed by fixed criteria. As captured by the notion of region building, this phenomenon can also appear as a project or goal pursued by political entrepreneurs who seek to enhance regional framing among industrialists, policy makers and others.[13] Neumann (1994: 59) may exaggerate the point when positing that regions are 'talked and written into existence', but the discursive aspect of regionality is to some extent separable from density of functional interaction. For instance, in the area now termed the Barents Region, the idea of the Northern hinterlands being joined together by their distinctiveness from the South was kept alive on the Nordic side by the North Calotte cooperation and widened geographically by the fact that the Pomor trade era was only a few generations past.[14]

The two faces of regionality will often stimulate one another: images and memories of historic ties are often mobilized to muster support for region-building efforts and to boost interaction among groups and

→ Carthreny et al (1994)

individuals within a region in the making.[15] Those who have pushed for greater cooperation under the Council of Baltic Sea States have frequently made references to the medieval Hansa trade networks that connected regional cities.[16] Often, those who favour an Arctic framing of problems and opportunities seek to counter-balance the predominant centre–periphery framing inherent in the fact that Arctic territories form segments of several larger countries whose centre of gravity and national priorities tend to lie elsewhere (Young 2000). Underlying our interest in region building is a concern about the conditions for peaceful change in the region. A more or less explicit objective of all the initiatives that were to give rise to new Arctic institutions was to facilitate the peaceful integration of Russia into cooperative structures – structures that had been wanting, or that had involved only Western states.

An obvious approach to studying evolving regionality is to examine changes in patterns of regional interaction at political levels or in the economic and cultural spheres. All the case studies in this book address such changes within their respective issue areas. However, the phenomena referred to by 'discursive distinctiveness' can be more difficult to measure. Such distinctiveness will sometimes be reflected in institutional features like separate bureaucratic bodies responsible for regional matters. Indeed, the establishment of circumpolar and subregional structures for cooperation is itself an indication that Northern territories are becoming recognized as distinctive in significant ways. It is equally clear that such formal-institutional regionality must have a substantive counterpart at the level of civil society if it is to remain a stable feature of Arctic affairs, in that regional-level solutions are regularly explored as relevant (if not always adequate) when responding to problems and opportunities. A political region requires a measure of common identity among state representatives, groups and individuals, and this must to a considerable extent be reflected in flows of interaction.

Conclusions

The analytical framework advanced in this chapter has the following components: Effectiveness is assessed by examining whether an institution, alone or interacting with other institutions, contributes significantly to the removal or mitigation of the problem that motivated its formation. Such contributions may occur by generating information about the severity of the problem or ways to mitigate it, by making international norms more compelling, or by altering the ability of relevant actors to behave in desirable ways or the cost associated with failure to do so. The same underlying causal mechanisms – shorthanded as 'cognitive', 'normative' or 'utilitarian' – are drawn upon when examining impacts on political participation in Arctic decision making and the development of closer ties between governments, organizations and individuals in the North. Such regional ties show up not only in direct interaction but also in how problems and opportunities are framed by players inside and outside the region.

In the following chapters, this framework structures the case studies of Arctic institutions at work in five important policy areas: indigenous affairs, communicable diseases, pollution control and biodiversity, climate change and environmental concerns in the oil and gas sectors.

Notes

1 These notions are elaborated below.
2 See Young (1999), Brown Weiss and Jacobson (1998), and Miles *et al.* (2002); trans-national effectiveness projects reported earlier include Haas *et al.* (1993) and Stokke and Vidas (1996).
3 The second of those mechanisms is essential to the 'logic of appropriateness' articulated by March and Olsen (1989: 21–6) the third to their 'logic of consequentiality', whereas the cognitive mechanism is an essential ingredient in both logics. My reason for treating cognitive influence as a separate mechanism is that improvement of factual knowledge about the problem at hand is among the most prominent activitities under international institutions. Note that a rather parallel trichotomy has emerged in the literature on policy instruments, which tends to distinguish between economic means, regulation, and information (Vedung 1998: 30–1).
4 See for instance Keohane *et al.* (1993: 7), Bernauer (1995: 364), Young and Levy (1999: 4–5), and Underdal (2002: 11).
5 See also Chapter 5 in this book.
6 See Underdal and Young (eds. 2004) and Oberthür and Gehring (eds. 2006). In focus in this book is horizontal interplay: i.e. that involving institutions at the same level of governance. Vertical interplay concerns the relationship between, for instance, international and national levels of governance; see Young (2002).
7 See for instance Brown Weiss (1993: 697–702) and Andresen (2001).
8 See in particular Young (ed. 1999), Stokke (ed. 2001), Miles *et al.* (2002), and Oberthür and Gehring (eds. 2006). This finding corresponds with earlier observations from single-regime evaluations: summarizing a set of case studes on environmental governance, Keohane *et al.* (1993: 15) found it 'somewhat surprising, but heartening, to discover that in our cases, cooperation among agencies is more salient than interinstitutional conflict'.
9 The negative version of this mechanism – sanctions in reaction to undesired behaviour – has not been relevant for Arctic institutions.

10 On the distinction between the normative and structural components of international regimes, see Stokke and Vidas (1996).
11 On the other hand, the Saami and the Nenets do not have a seat in the intergovernmental Barents Council.
12 The clearest evidence of this linkage is found in the Joint Declaration from the Meeting of the Ministers of Environment of the Nordic Countries and the Russian Federation, Kirkenes 3–4 September 1992, which states that 'The Ministers recognize that solving the existing major trans-boundary environmental problems will be central in realizing the potential for broader cooperation in the Barents Region. . .'.
13 On the distinction between (top–down) 'regionalism' and (bottom–up) 'regionalization', and the attempt to transcend that distinction by a 'region-building' approach, see for instance Neumann (1994) and Keskitalo (2004: 6–11).
14 On the Russian side, this would be true for Arkhangelsk oblast in particular. Unlike the inhabitants of Murmansk, who moved into the region only after the Russian Revolution, many in Arkhangelsk still use the Pomor era as an important frame of reference in judging the behaviour of foreigners (Castberg *et al.* 1994: 72). The Pomor trade between the northerly parts of Russia and Norway lasted for nearly two centuries up till the Russian Revolution, and mainly involved the exchange of Norwegian fish for Russian grain and wood products from the area around Arkhangelsk For a presentation of the North Calotte partnership, see www.nordkalottradet.nu.
15 See for instance Keskitalo (2004).
16 On this *Mare Balticum* rhetoric, see Chapter 4 by Hønneland and Rowe, this volume.

References

Andresen, Steinar (2001) 'Global environmental governance: UN fragmentation and co-ordination', in O. S. Stokke and Ø. B. Thommessen (eds.), *Yearbook of International Co-operation on Environment and Development 2001/2002*, London: Earthscan, pp. 19–26.

Bernauer, Thomas (1995) 'The effect of international environmental institutions: how we might learn more', *International Organization*, 49: 351–77.

Bodansky, Daniel (1999) 'The legitimacy of international governance: a coming challenge for international environmental law? *American Journal of International Law*, 93: 596–624.

Brown Weiss, Edith (1993) 'International environmental law: contemporary issues and the emergence of a new world order', *Georgetown Law Journal*, 81: 675–710.

Brown Weiss, Edith and Harold K. Jacobson (eds.) (1993) *Engaging Countries: Strengthening Compliance with Environmental Accords*, Cambridge, MA: MIT Press.

Castberg, Rune, Stokke, Olav Schram and Østreng, Willy (1994) 'The dynam-

the Barents Region' in O. S. Stokke and O. Tunander (eds.), *The ts Region: Cooperation in Arctic Europe*, London: SAGE, pp. 71–84.

Thomas M. (1990) *The Power of Legitimacy Among Nations*, New : Oxford University Press.

George, Alexander L. and McKeown, Timothy J. (1985) 'Case studies and theories of organizational decision making', *Advances in Information Processing in Organizations*, 2: 21–58.

Haas, Peter M., Keohane, Robert O. and Levy, Marc A. (eds.) (1993) *Institutions for the Earth: Sources of Effective International Environmental Protection*, Cambridge, MA: MIT Press.

Haavisto, Pekka (2001) *Review of the Arctic Council Structures*, Helsinki: Finnish Institute of International Affairs.

Hannan, Michael T. and Freeman, John (1977) 'The population ecology of organizations', *American Journal of Sociology*, 82: 929–64.

Joint Declaration from the Meeting of the Ministers of the Environment of the Nordic Countries and the Russian Federation (held in Kirkenes, Norway, 3–4 September 1992), available from the Norwegian Ministry of the Environment, Oslo.

Kay, David A. and Jacobson, Harold K. (eds.) (1983) *Environmental Protection: The International Dimension*, Totowa, NJ: Allanheld, Osmund & Co. Published under the auspices of American Society for International Law.

Keohane, Robert O. (1984) *After Hegemony: Cooperation and Discord in the World Political Economy*, Princeton, NJ: Princeton University Press.

Keohane, Robert O., Haas, Peter M. and Levy, Marc A. (1993) 'The effectiveness of international environmental institutions' in P. M. Haas, M. A. Levy and R. O. Keohane (eds.), *Institutions for the Earth: Sources of Effective International Environmental Protection*, Cambridge, MA: MIT University Press, pp. 3–26.

Keskitalo, E. C. H. (2004) *Negotiating the Arctic: The Construction of an International Region*, New York: Routledge.

Krasner, Stephen D. (ed.) (1983) *International Regimes*, Ithaca, NY: Cornell University Press.

March, James G. and Olsen, Johan P. (1989), *Rediscovering Institutions: The Organizational Basis of Politics*, New York: The Free Press.

Mearsheimer, John J. (1995) 'The false promise of international institutions', *International Security*, 19: 5–49.

Miles, Edward L., Underdal, Arild, Andresen, Steinar, Wettestad, Jørgen, Skjærseth, Jon Birger and Carlin, Elaine M. (2002) *Environmental Regime Effectiveness: Confronting Theory with Evidence*, Cambridge, MA: MIT Press.

Neumann, Iver B. (1994) 'A region-building approach to Northern Europe', *Review of International Studies*, 20: 53–74.

Oberthür, Sebastian and Gehring, Thomas (eds.) (2006) *Institutional Interaction in Global Environmental Governance: Synergy and Conflict among International and EU Policies*, Cambridge, MA: MIT Press.

Runge, C. Ford (1994) *Freer Trade, Protected Environment: Balancing Trade Liberalization and Environmental Interests*, New York: Council on Foreign Relations.

Schoenbaum, Thomas J. (1997) 'International trade and protection of the environment: the continuing search for reconciliation', *American Journal of International Law*, 91: 268–313.

Stokke, Olav Schram (1994) 'Environmental cooperation as a driving force in the Barents Region' in O. S. Stokke and O. Tunander (eds.), *The Barents Region: Cooperation in Arctic Europe*, London: SAGE, pp. 145–159.

Stokke, Olav Schram (1998) 'Nuclear dumping in Arctic seas: Russian implementation of the London Convention' in D. G. Victor, K. Raustiala and E. B. Skolnikoff (eds.), *The Implementation and Effectiveness of International Environmental Commitments: Theory and Practice*, Cambridge, MA: MIT Press, pp. 475–517.

Stokke, Olav Schram (2004) 'Boolean analysis, mechanisms, and the study of regime effectiveness' in A. Underdal and O. R. Young (eds.), *Regime Consequences: Methodological Challenges and Research Strategies*, Dordrecht: Kluwer Academic, pp. 87–119.

Stokke, Olav Schram (ed.) (2001) *Governing High Seas Fisheries: The Interplay of Global and Regional Regimes*, Oxford: Oxford University Press.

Stokke, Olav Schram and Davor Vidas (1996) 'The effectiveness and legitimacy of international regimes' in O. S. Stokke and D. Vidas (eds.), *Governing the Antarctic: The Effectiveness and Legitimacy of the Antarctic Treaty System*, Cambridge: Cambridge University Press, pp. 13–31.

Stokke, Olav Schram and Davor Vidas (eds.) (1996) *Governing the Antarctic: The Effectiveness and Legitimacy of the Antarctic Treaty System*, Cambridge: Cambridge University Press.

Thelen, Kathleen and Steinmo, Sven (1992) 'Historical institutionalism in comparative politics' in S. Steinmo, K. Thelen and F. Longstreth (eds.), *Structuring Politics: Historical Institutionalism in Comparative Analysis*, Cambridge: Cambridge University Press, pp. 1–32.

Thompson, William R. (1973) 'The regional subsystem: a conceptual explication and a propositional inventory', *International Studies Quarterly*, 17: 89–117.

Underdal, Arild (1992) 'The concept of regime "effectiveness"', *Cooperation and Conflict*, 27: 227–40.

Underdal, Arild (2002) 'One question, two answers' in E. L. Miles, A. Underdal, S. Andresen, J. Wettestad, J. B. Skjærseth and E. M. Carlin, *Environmental Regime Effectiveness: Confronting Theory with Evidence*, Cambridge, MA: MIT Press, pp. 3–45.

Underdal, Arild and Young, Oran R. (eds.) (2004). *Regime Consequences: Methodological Challenges and Research Strategies*. Dordrecht: Kluwer Academic.

Vedung, Evert (1998) 'Policy instruments: typologies and theories' in M.-L. Bemelmans-Videc, R. C. Rist, and E. Vedung (eds.), *Carrots, Sticks, and*

Sermons: Policy Instruments and their Evaluation, New Brunswick, NJ: Transaction Publishers, pp. 21–58.

Veggeland, Noralv (1994) 'The Barents Region as a European frontier region' in O. S. Stokke and O. Tunander (eds.), *The Barents Region: Cooperation in Arctic Europe,* London: SAGE, pp. 201–12.

Yin, Robert K. (1989) *Case Study Research: Design and Methods.* London: SAGE (Applied Social Research Methods Series, Volume 5).

Young, Oran R. (2000) 'The structure of Arctic cooperation: solving problems/ seizing opportunities', unpublished paper prepared at the request of Finland for the Fourth Conference of Parliamentarians of the Arctic Region, Rovaniemi, 27–29 August 2000, and the Finnish chairmanship of the Arctic Council 2000–2002.

Young, Oran R. (2002) *The Institutional Dimensions of Environmental Change: Fit, Interplay, and Scale,* Cambridge, MA: MIT Press.

Young, Oran R. (ed.) (1999) *The Effectiveness of International Environmental Regimes: Causal Connections and Behavioral Mechanisms,* Cambridge, MA: MIT Press.

Young, Oran R. and Levy, Marc A. (with Gail Osherenko) (1999) 'The effectiveness of international regimes' in O. R. Young (ed.), *The Effectiveness of International Environmental Regimes: Causal Connections and Behavioral Mechanisms,* Cambridge, MA: MIT Press, pp. 1–33.

3 Indigenous issues

Elana Wilson and Indra Øverland

Introduction

This chapter examines the impact of Arctic regimes on indigenous is-sues.[1] The focus is on the Arctic Council and the Barents Euro-Arctic Region (BEAR). Other Arctic regimes and global indigenous regimes are discussed only briefly, in order to locate the Arctic Council and the BEAR in the wider landscape of international cooperation.

Out of a circumpolar population of approximately four million peo-ple (Bogoyavlenskiy and Siggner 2004), it has been estimated that over 320,000 are indigenous (NOU 2003). The Arctic Council officially recognizes 24 Arctic indigenous peoples, but this is a low figure (NOU 2003). In Russia alone, 39 peoples are recognized as 'numerically small indigenous peoples of the Russian North, Siberia and the Far East' – and that does not include peoples numbering more than 50,000, such as the Komi or the Yakut.[2] According to the International Labour Organization (ILO) Convention 169 (ILO 1989), those peoples should also be recognized as indigenous.

The Arctic indigenous peoples face a range of challenges and prob-lems, most of which they share with indigenous peoples elsewhere: threatened or unacknowledged rights to land and nature resources; (relative) poverty; separation by state borders; environmental prob-lems like persistent organic pollutants (POPs), oil spills and climate change; disenfranchisement and lack of representation in national and international politics.

Unrecognized or poorly implemented rights to land and natural resources may be the greatest problem for indigenous peoples in most of the world (Wiben Jensen 2004), including the Nordic countries and Russia, though decreasingly so in Canada and the USA. The lack of recognized rights is particularly damaging in the face of industrial developments that directly or indirectly impact on indigenous lands (see European Environment Agency 2004). In the Arctic context, the question of land rights is de facto a domestic issue not addressed by the international Arctic regimes. Furthermore, such land issues are often connected to oil and gas developments – which have also been defined out of the Arctic Council's responsibilities. It is therefore not possible to assess the Arctic indigenous regimes' response to this problem, as one might in the case of global indigenous regimes like ILO Convention 169, which attempts to set minimum standards for the recognition of indigenous land rights by various states.

It is perhaps not incidental that land rights are not addressed in Arctic regimes. Finland was one of the first states to sign ILO Convention 169, but still had not ratified it as of December 2005, due to uncertainties about the implications for land ownership in the northerly Finnish province of Lapland (Toivanen 2002: 7–8). Similarly, Canada, Russia, Sweden and the USA had not ratified the convention as of December 2005. Norway was the first European country to ratify ILO Convention 169, but has been criticized by Norwegian Saami, who felt that implementation of the convention was badly planned and that the clauses on land ownership were interpreted in a restrictive way (Toivanen 2002). Thus, the Arctic states have not been at the forefront of the main global indigenous regime, so it is hardly surprising that Arctic regimes have been limited in their ambitions on specifically indigenous issues.

Nonetheless, the indigenous theme has permeated discourses within the Arctic Council and the BEAR (Fjellheim 2003). It is also important to note that key issues taken up by the Arctic Council – such as transboundary pollutants, sustainable development and climate change –are considered to be fundamental indigenous problems by all indigenous organizations involved. However, the problem specific to indigenous peoples that has received the greatest attention in the international Arctic regimes is that of political representation, and this has been addressed by providing for indigenous representation through the regimes themselves.

Beyond this shared attention to the question of representation, indigenous problems have been addressed in the Arctic Council and the BEAR in divergent ways. In the Arctic Council, indigenous organizations have utilized their relatively strong position within the Council to influence states, and have used the scientific and policy documents produced by Arctic Council working groups as foundations for their advocacy efforts in several international arenas. In the BEAR, the primary focus has been on aid and development projects geared towards increasing regional interaction and reducing poverty in Russia.

Consequently, in this chapter we focus on the roles that political representation, support for Russian indigenous peoples and their organizations and the production of scientific studies have played in addressing indigenous issues and facilitating political mobilization and region building in the Arctic. These three areas of activity are often dealt with in an integrated fashion or by shifting between regimes, with a particular focus on the extent to which indigenous organizations have been able to utilize their representation within both the Arctic Council and BEAR to pursue their own agendas.

Regime-based responses

The Arctic Council

Six indigenous organizations now participate as non-voting 'permanent participants' at the Arctic Council and in working groups. According to the founding document of the Arctic Council, the permanent participant category is meant to 'provide for active participation and full consultation with the Arctic indigenous representatives within the Arctic Council' – with the caveat that permanent participants cannot outnumber member states (Arctic Council 1996).

This increase in indigenous representation, however, was by no means a certain outcome in the negotiations leading up the establishment of the Arctic Council. The Canadian delegation worked extensively with leaders of the Inuit Circumpolar Conference (ICC) to lobby for meaningful indigenous participation in the Arctic Council (Keskitalo 2004; Tennberg 2000). Russia seemed to agree to the idea of indigenous participation with few reservations, while the USA was more reluctant (Keskitalo 2004; Tennberg 2000). The US opposition to the Canadian focus on indigenous issues and representation was due in part to a strong environmental protectionist lobby whose interests had often been at odds with indigenous interests. Consequently, US delegates 'saw proposals related to indigenous rights as an [opening] for attacking US marine mammal [protection] legislation' (Keskitalo 2004: 72).[3] The disagreement over indigenous participation in the Arctic Council remained an issue through June 1996, when the US head of the delegation required that the founding document of the Arctic Council omit a preamble on environmental security and indigenous peoples and avoid the use of the term 'peoples', which would imply that indigenous groups hold special, collective rights (Keskitalo 2004).

Five of these six permanent participant organizations represent indigenous peoples whose populations spread across state boundaries.

The ICC (Canada, USA, Greenland, Russia), the Saami Council (Norway, Sweden, Finland, Russia) and RAIPON (the Russian Association of Indigenous Peoples of the North) were the three initial organizations to advocate for and gain permanent participant status within the Arctic Council. These groups proceeded to set the terms for the accession of other permanent participants to the Arctic Council, namely that the organizations must represent one indigenous people living in more than one Arctic state, or several indigenous peoples from one state (Keskitalo 2004). In 2000, three more organizations became permanent participants: the Arctic Athabaskan Council (Canada, USA), the Gwich'in International Council (Canada, US) and the Aleut International Association (USA, Russia).

The participation of indigenous organizations and their representatives in Arctic Council negotiations and events is partly supported by the Indigenous Peoples Secretariat (IPS), which was established in 1994 to encourage and support the participation of indigenous organizations in the Arctic Environmental Protection Strategy (the precursor of the Arctic Council). The IPS has been funded primarily, although not exclusively, by the government of Denmark (IPS 2005).

The Barents Euro-Arctic Region (BEAR)

The *Kirkenes Declaration*, the founding document of the BEAR, includes a section dedicated to indigenous peoples. In this section, 'the participants concerned reaffirmed their commitment to the rights of the indigenous peoples in the North in keeping with the objectives set out in chapter 26 of Agenda 21', 'took note' of the proposal to establish a Working Group for Indigenous Issues, mentioned various possible projects and activities related to indigenous peoples and 'agreed to exchange information regarding existing or proposed legislation with a bearing on the position of indigenous peoples in their respective countries'.

The use of the formulation 'the participants concerned. . .' is peculiar. It is not used elsewhere in the Kirkenes Declaration, and it is not clear why it is used in the section on indigenous peoples or exactly what it means. It may indicate that not all participants were equally interested in including indigenous issues in the remit of the Barents institutions.

The highest organ of the BEAR, the foreign ministerial Barents Council, has no indigenous representation. The other major Barents institution, the Regional Council, includes one member who represents the three peoples in the region recognized as indigenous by the governments of the states in which they reside – the Saami, the Nenets and the Veps.

In addition, a working group on indigenous issues has been formed as part of the BEAR, including representatives of the indigenous peoples. The working group has an independent advisory function in relation to the Barents Council and the Regional Council and is thus a potentially influential organ. Also, its title has been altered from 'Working Group *for* Indigenous Issues', the formulation used in the Kirkenes Declaration, to 'Working Group *of* Indigenous Peoples', which could be taken to imply the strengthening of the representational aspect of the working group. It is also the only working group within the BEAR that has been in continuous existence since the mid-1990s. Its level of activity and impact, however, have suffered from a lack of resources. In a strongly worded press release from 2004, the working group laments the 'patronising and discriminating attitude' towards indigenous involvement in the BEAR as indicated by the lack of will to provide the financial support necessary for the working group to function (Nilsen 2004).

Other regional regimes

Despite the proliferation of regional organizations in the Nordic region, there is little overlap on indigenous issues. The oldest of the regional organizations, the Nordic Council and Nordic Council of Ministers, include no indigenous representation (Fjellheim 2003). The list of the fields of cooperation of the two institutions includes 27 different areas – for example, adult education, fisheries, gender equality, information technology and drug prevention – but not indigenous issues (Nordic Council 2005). Thus, although the Nordic institutions have supported several projects related to indigenous peoples, such as the Nordic Saami Institute in Kautokeino, indigenous issues are not an important focus for their activities, nor are there rigorous mechanisms for indigenous representation in their own work. Considering that every single member of the Nordic institutions has a significant indigenous population, this is somewhat surprising. The Council of the Baltic Sea States, which includes most of the countries that are part of the BEAR and the Nordic Council and Nordic Council of Ministers, has a similarly low profile on indigenous issues. The Conference of Parliamentarians of the Arctic Region has a secretariat in Reykjavik and between its biannual conferences is run by the 13-person Standing Committee that includes three indigenous representatives. The Conference of Parliamentarians of the Arctic encourages states and international organizations to promote indigenous representation and to take indigenous issues seriously. However, its main practical significance is as a clearinghouse for initiatives aimed at the Arctic Council, in which the Standing Committee has observer status.

Effectiveness

In assessing the effectiveness of the Arctic Council and the BEAR on indigenous issues, the impacts and efficacy of indigenous political representation, scientific knowledge and aid projects on indigenous problems will be taken as yardsticks. The effectiveness of indigenous representation and of information produced about the condition of the Arctic environment are examined through two case studies of the ICC's political efforts in which the Arctic Council served both as a forum and as a source of information. The efficacy of aid and development projects is examined in relationship to the activities of the BEAR.

Political representation and environmental knowledge in the Arctic Council: The ICC, POPs and climate change

The ICC is well positioned to take advantage of representation at the Arctic Council. The organization was founded in 1977 to represent Inuit in the United States (Alaska), Canada (the Northwest Territories, Nunavut, Nunavik-Northern Quebec and Labrador), Greenland and Russia (Chukotka). It has since grown into a large and experienced organization representing 150,000 Inuit at the international level. Through several international forums, the ICC has succeeded in framing environmental problems as human rights issues for indigenous peoples and in promoting awareness of indigenous peoples' capability to manage their own resources sustainably and successfully (Innukshuk 1994; Nuttall, 1998; Tennberg 2000). Young (1992: 208) argues that the ICC has gained 'a certain moral standing as the voice of the Arctic's permanent residents'.

The issues pursued by the ICC in the case studies that follow – transboundary pollutants and climate change – clearly fall within the mandate and agenda of the Arctic Council. The analysis of a reputable and efficient organization working within a well-accepted Arctic issue area serves to illuminate the possibilities of and limitations to the effective use of information produced by the Arctic Council and of indigenous representation within it. This study may indicate the opportunities and challenges that other indigenous organizations, operating at the Arctic

regional level and discussed in the subsequent section on political mobilization, will face as they continue to develop.[4]

As to the representation yardstick, there are several underlying
questions. Given that permanent participants on the Arctic Council
are not allowed to vote, the gauge of effectiveness used here is based
on the concept of influence. Has the ICC, due to the existence of the
Arctic Council, been able to more successfully access and influence
states that can co-champion or support indigenous political causes? In
terms of assessing the effectiveness of policy statements and scientific
studies about the state of the Arctic environment, we will examine the
extent to which ICC leaders have been able to use these documents to
influence key actors and support their positions in domestic, Arctic and
international arenas.

The work of the ICC on the issue of persistent organic pollutants
(POPs) clearly demonstrates the need for indigenous organizations to
utilize connections and capitalize on possibilities in a range of institutions and forums, including the Arctic Council, in order to achieve
desired ends. The question of persistent organic pollutants appeared
on the international political agenda in the 1990s and resulted in two
international binding treaties to eliminate certain POPs – the Århus
Protocol on Persistent Organic Pollutants in 1998 (based on the 1979
Convention on Long-Range Transboundary Air Pollution) and the
Stockholm Convention on Persistent Organic Pollutants in 2001.[5] These
protocols were in part fuelled by the publication of the findings of the
Canadian Northern Contaminants Program and the Arctic Monitoring
and Assessment Programme (AMAP) (part of the Arctic Environmental
Protection Strategy and later the Arctic Council), which documented
how contaminants travelled from sources in the temperate world and
into the Arctic environment, its wildlife and residents (Downie and
Fenge 2003).

Northern Canadian indigenous leaders sought to ensure that POPs
were seen as both a public health issue and as a threat to the human
rights of indigenous peoples (Downie and Fenge 2003; *Nunatsiaq News*
1998). Indigenous organizations did not easily find a niche within the
ongoing POPs negotiations held primarily under the auspices of the
UN Environment Programme (UNEP), and the ICC's cooperation with
Canadian governmental officials was far from straightforward.[6] While
working to develop an effective relationship with the Canadian government, the ICC, along with RAIPON and the Saami Council, gained observer status in the AEPS and made participation in the AMAP working
group a priority. The report released by AMAP inspired AEPS ministers
to commit themselves to making 'a determined effort to secure support

for international action which will reduce Arctic contamination' (Arctic Council 1997: 1). The AEPS and the Arctic Council were essential in promoting international action for two reasons. First, the extensive and Arctic-specific information that had been collected, assembled and verified by state representatives under AEPS provided a crucial knowledge base. Secondly, indigenous peoples were able to use their 'permanent participant status to prod the Arctic states to international action' (Fenge 2003: 211).

Furthermore, the Arctic Council became an important arena in which to gain the attention and support of the USA on the POPs issue. Until a 1998 Arctic Council meeting held in Barrow, Alaska, US officials did not acknowledge the Arctic dimension of the POPs issue – that these pollutants were concentrating in the North. At this meeting, however, Alaskan interests, including the Alaskan governor and ICC-Alaska representatives, had the opportunity to push American delegates to take a more positive stance towards the POPs negotiation process (Fenge 2003; Watt-Cloutier 2003).

This brief case study indicates two central findings in terms of the effectiveness of the ICC in using both Arctic Council representation and information to pursue international conventions on the elimination of POPs. First, much of the ICC's success in influencing POPs negotiations was due to the mutually helpful relationship they fostered with key persons and departments within the Canadian federal government and through directly lobbying Arctic governments via the Arctic Council. It is worth noting here, however, that a similar relationship between an indigenous organization and big powers like the USA or Russia may be unlikely. Indigenous issues occupy a much lower place on the federal agenda of such states, so federal departments and relevant politicians and civil servants may not be as readily influenced and accessed by indigenous organizations as they are in Canada. Furthermore, Fenge (2003) notes that even in Canada, where the ICC had already cultivated important Canadian governmental contacts through work within the AEPS and other Arctic processes, the spirit and practice of cooperation did not automatically extend into the new field of POPs negotiations, as the process involved different departments and state actors. Thus, influence built up by an indigenous organization within one particular Arctic regime – in this case the Arctic Council – may not automatically translate into new departments and issue areas, even under the best of circumstances.

Second, the Arctic Council was an essential source of information. The AMAP report provided a crucial foundation of widely verified and endorsed information about the physical state of the Arctic environment upon which the ICC could build their case. This type of extensive and accepted scientific information would be impossible for one state, let alone a non-profit indigenous organization, to produce independently.

Information released by the Arctic Council has been decisive for ICC efforts in another key cause – drawing attention to and advocating for swift mitigation of climate change. The Arctic Council recently released the findings of the Arctic Climate Impact Assessment (ACIA) on the effects of climate change on the ecosystems and communities of the Arctic (discussed further by Hoel in this volume). The ACIA findings project impacts on vegetation and habitats, on coastal communities, sea ice coverage, permafrost and the infrastructure built upon it, indigenous economies and communities and UV levels. The ACIA Policy Document, endorsed by the eight member states of the Arctic Council and the six permanent participants, outlines a general commitment on the part of member states to work further on the issue of climate change in both research and policy (Arctic Council 2004).

ICC Chair Sheila Watt-Cloutier has frequently drawn upon ACIA research and the endorsement of ACIA findings by the Arctic Council to support her arguments (see Watt-Cloutier 2005a,b). However, the ICC has also had to pursue other avenues outside the Arctic Council in order to exert influence over the USA. The most prominent of these non-Arctic efforts is an ICC petition to the Inter-American Human Rights Commission system.

ICC leaders claim that climate change is a human rights issue for Inuit. In a recent interview, Watt-Cloutier put it this way: 'It's not just about the environment – it's about culture. It's about a way of life, all of those things which are so important to us in the Arctic' (in George 2005). The petition, which was released on 7 December 2005, declares:

> Protecting human rights is the most fundamental responsibility of civilized nations. Because climate change is threatening the lives, health, culture and livelihoods of the Inuit, it is the responsibility of the United States, as the largest source of greenhouse gases, to take immediate and effective action to protect the rights of Inuit.
>
> (Watt-Cloutier 2005c: 7)

In the case of climate change, the importance of scientific information assembled and interpreted through the Arctic Council is indisputable. However, the limits of the Arctic Council also become evident. The ICC clearly wants to affect the policy of the USA – one of the major polluters and a great power. However, ICC leaders have chosen to pursue this goal outside of the Arctic Council, which indicates that their ability to influence Washington within the Arctic Council is limited. It would thus appear that indigenous organizations may be able to use their Arctic Council representation for effective lobbying of states with extensive and specifically Northern interests, but must present their case elsewhere if they are to gain the attention of greater powers defending national interests.

Aid and development projects under the BEAR

In practice, the most noticeable activity in the BEAR has been the funding of projects, most of which have been bilateral projects funded by Norway (Foreign Affairs Committee 2001: 3), especially oriented towards aid for people and organizations in Russia.[7] The indigenous dimension is no exception in this regard. An overview of projects during the years 2002 and 2003 shows that 43 were financed by Norway, nine by Finland and three by Sweden. Judging from the project titles, at least 24 were clearly oriented towards Russia and only four were clearly not oriented towards Russia (Prakhova *et al.*, 2004). Most projects have been carried out under the auspices of the Norwegian Barents Secretariat in Kirkenes. This exemplifies a broader trend in which the Barents Secretariat has come to assume such a central position in the BEAR that what was initially conceived as a platform for multinational projects has in fact become a channel for bilateral Norwegian aid to Russia (see NOU 2003).

Several sources have lauded the focus on small-scale projects in the BEAR as a success (e.g. Nystad 2003; Landsdelsutvalget 2005). They are seen as more useful, and therefore the Barents institutions as more important and worthwhile, than the Arctic Council (Offerdal 2003). It is, however, difficult to verify such views, as the multi-level and fragmented nature of the Barents institutions in itself hinders an overall assessment its results.

One project on which relatively detailed information could be obtained – the Choom National Cultural Centre in the Russian Saami 'capital' of Lovozero – exemplifies another problem. An internal note from the Barents Secretariat in Kirkenes indicates that at least 2.9 million NOK (approximately GBP 251,000) has been spent on the project. In practice this funding has gone to feasibility studies, minor improvements of the façade of the building, painting the interior, replacing windows and installing a new heating system and similar efforts. Although a small part of the sum has also been spent on feasibility studies for a similar Nenets national cultural centre, the funding allocated surpasses the realistic cost of such an undertaking in Russia. The project also resulted in an open conflict when the employees of the Lovozerstroi company went on strike for non-payment of wages and the various Norwegian and Russian actors involved started blaming each other for failing to deliver promised financing. Of course the Choom project is not representative of all BEAR indigenous projects, but it may indicate some important pitfalls. In addition, it is of some importance as a high-profile project in the history of BEAR support for indigenous activities, both because of the volume of funding involved and because it was mentioned in the Kirkenes Declaration and is physically visible in the Russian Saami landscape.

On a more positive note, the many Barents indigenous projects have undoubtedly strengthened contacts between the Saami in the Nordic countries and the Saami and other indigenous peoples in northwestern Russia. In total, more than 2000 Norwegian–Russian projects were carried out by the Barents Secretariat in Kirkenes from 1993 to 2004 (MFA 2005). Extrapolating from Prakhova, Nystad and Mikhalyuk's (2004: 5–6) figures for 2002 and 2003, we can estimate that at least one quarter of these targeted indigenous peoples. With so many projects, there is little doubt that the indigenous peoples of northwestern Russia have benefited from the creation of the BEAR.

Political mobilization

Arctic regimes have resulted in producing both new actors and in changing, to some extent, the patterns of interaction and influence in decision making on Arctic affairs. In the case of the Arctic Council, the representation of indigenous organizations has clearly had an effect on political mobilization. First and foremost, the establishment of high-level Arctic political forums served as an impetus for indigenous peoples with primarily domestic organizations to mobilize more effectively for participation at the international level. This point is substantiated by the establishment of the Arctic Athabaskan Council[8] and the Gwich'in Council International[9] and their accession to the Arctic Council as permanent participants in 2000 (IWGIA 2001). Although the Gwich'in and Athabaskan peoples already had trans-boundary relations with their co-ethnics in Alaska, these new organizations represent their response to the offer of representation at the circumpolar level. Another later addition to the Arctic Council is the Aleut International Association. This organization presents its history and development in a longer-term historical perspective, beginning with increased Aleut interaction across the Bering Sea during *perestroika* in the late 1980s. However, the realization of an actual organization in 1998 indicates that the existence of the Arctic Council was an important push factor for formalizing these relations.[10]

Furthermore, in the case of the Arctic Council, the category of 'permanent participant' gave more established indigenous organizations, like the ICC and the Saami Council, the motivation and context to engage in capacity-building work with the more nascent indigenous movement in Russia. Indeed, Murashko (2002: 26) comments: 'RAIPON has improved its political and executive structure primarily through funding from international projects'. One such initiative, the Institution Building for Northern Russia's Indigenous Peoples Project (INRIPP), was funded by the Canadian International Development Agency and implementedt by the Department of Indian and Northern Development (Government of Canada) and the ICC. The project focused on lending practical assistance, such as the distribution of office and communications equipment to thirty regional offices of RAIPON, the establishment of a central office for RAIPON in Moscow and the facilitation and funding of RAIPON's participation in the Arctic Council.[11] ICC and Canadian governmental representatives had lobbied for the meaningful participation of indigenous peoples in the Arctic Council and thus had an interest in ensuring that there was an effective organization to represent Russian indigenous peoples to reinforce the position of permanent participants on the Council (Wilson 2006). These efforts could, alternatively, be interpreted as part of the ICC's focus on international development worldwide, or as part of a sense of solidarity with and obligation to indigenous peoples worldwide (and the Inuit of Chukotka in particular). However, the close chronological relationship between the establishment of INRIPP (1996) and the founding of the Arctic Council suggests that entrenching the category of 'Permanent Participant' was a significant motivation behind the assistance to the Russian indigenous movement.

While BEAR projects have not resulted in the production or support of new actors, the vast number of small-scale projects aimed at solving local problems has generated numerous new cross-border contacts. The extent to which these contacts have supported the efforts of emerging indigenous organizations amongst the Kola Saami is difficult to document. However, it is plausible that these connections may serve as sources of information, funding or advice. In a more critical perspective, one might ask how many of these projects include significant elements of what might be called 'aid tourism'. For example, two of the projects in the 2002–2003 overview are listed as 'Excursion to the Kola Peninsula' and 'Study Trip to Lovozero' (Prakhova *et al.*, 2004). Furthermore, it is notable that most of the traffic created by the flourishing portfolio of Barents projects involves movement from the West, in particular Norway and Finland, to the East – i.e. Russia. During some periods at least two busloads of foreigners would arrive in Lovozero every week.

In addition to bringing new indigenous actors to the circumpolar scene, Arctic regimes may represent arenas for forging new relationships between states and indigenous organizations. In the ICC case studies outlined above, the Arctic Council was a key forum in which the ICC and the Canadian delegation could build up a mutually beneficial partnership. Furthermore, in negotiations leading up to the establishment of the Arctic Council, even the US delegation pressed (unsuccessfully) for the addition of a US-based indigenous group to represent non-Inuit Native Alaskans (Keskitalo 2004; Tennberg 2000). It is evident that the prominent position of indigenous organizations in the Arctic Council creates a situation in which it is advantageous for state representatives to develop and maintain indigenous allies. This represents a partial reversal of the more typical political scenario in which indigenous organizations have to work hard to influence states.

The more limited indigenous representation provided for in the BEAR has inevitably had less impact on specifically political mobilization for indigenous issues. Important obstacles in this respect are both the lack of direct indigenous representation at the highest level of the BEAR – the Barents Council – and the lack of funding for the preparation of input by the Working Group of Indigenous Peoples. The fact that indigenous representatives are not influentially situated within BEAR reduces the incentive of voting members to build partnerships and alliances with indigenous organizations. This reduces the impact of the BEAR on promoting new types of interaction and problem-solving with indigenous organizations or around indigenous issues.

Impacts on region building

In considering the impacts of the Arctic regimes on region building we also need to take into account other drivers and processes that may contribute to region building amongst indigenous peoples in the Arctic. The activities and development projects outlined above point to increased regional interaction between indigenous peoples and organizations. However, regional interaction between Northern indigenous organizations and leaders predates both the Arctic Council and BEAR by several decades.[12] In this section, we position this regional interaction within the broader history of indigenous trans-nationalism in the Arctic. We will also examine how Arctic regimes have fostered and sustained this interaction and the extent to which indigenous peoples have worked to present the Arctic as a distinctive regional unit.

Indigenous peoples and their organizations have promoted the notion of the Arctic as a region as a way of locating their concerns clearly on the map. Young (1992: 187) notes that indigenous peoples and such organizations as the ICC and the Saami Council had 'taken the lead in promoting international cooperation and awareness of the Arctic as a distinct region'. The ICC, for example, frequently works to have the Arctic region recognized as a vulnerable region of concern in international documents and declarations and as a barometer for global health. Watt-Cloutier (2005a), in a speech to the United Nations Environment Programme, argued: 'What we are experiencing today you will experience tomorrow. The Arctic is the world's climate change barometer, and Inuit are the mercury in that barometer.' On the other hand, the indigenous organizations' Arctic regional emphasis and region-building activities, like the efforts of the ICC and Saami Council in Russia, are intertwined with other developments at the national and regional levels, many of which predate the Arctic regimes addressed in this volume.

Firstly, the increased politicization of Arctic indigenous peoples at the domestic, regional and international levels has been related to the development of the North as a resource region and to the attendant focus of national governments on asserting sovereignty over the North and extracting resource wealth from it. The drive towards resource extraction brought an influx of outsiders and raised indigenous concerns about the protection of the environment and the right of the indigenous people to participate in and benefit from development, or prevent development from taking place on their lands (Young 1992; Mitchell 1996). For example, Mitchell (1996) argues that the ICC was initially conceived in 1977 in response to a request made by the Inuvialuit of the Beaufort Delta (Northwest Territories, Canada) to Inuit of the North Slope Borough in Alaska to support the Inuvialuit in their efforts to prevent offshore drilling in the Beaufort Sea.

Secondly, indigenous international advocacy at the Arctic regional and international levels has also been a result of the trans-national activities of *other* organizations. In the Canadian Arctic, for example, Inuit organized so as to respond to the trans-national activities of animal protectionists working to curtail or even eliminate some types of marine mammal hunting that are integral to Inuit culture and economy (Innukshuk 1994; Young 1992).

Thirdly, the growth of understanding of the Arctic region and of partnership between Arctic indigenous organizations and peoples can be seen as a product of the Fourth World movement and the globalization of indigenous politics. For example, the Arctic Peoples Conference in Copenhagen (1973) is seen as the beginning of 'modern indigenous internationalism' (Jull 1999). Since then, other forums for international indigenous activism and negotiating indigenous identity and politics have emerged, most notably the UN Working Group on Indigenous Populations. It was through indigenous international action that the concept of 'indigenous' and 'indigeneity' gained political currency (Nuttall 1998; Maaka and Fleras 2000; Muehlebach 2003).

Furthermore, in the case of the BEAR, we must take into consideration one further development with roots outside Arctic regimes – the dissolution of the Soviet Union and the opening of the Russian border. This makes it difficult to say to what extent it is the BEAR that has promoted more frequent cross-border contacts and a stronger sense of regional identity, and to what extent these interactions may have increased of their own accord due to more open borders in the European North.

The preponderance of Norwegian-Russian aid projects in the BEAR may also have had a special impact on prospects for region building. Despite the heady rhetoric on regional unity, most projects oriented towards indigenous peoples are often centred on Norwegian and Russian partnerships, and may thus be more a case of bilateral relations than regional identities. Furthermore, because the projects are so often focused on Western aid to Russian indigenous peoples, they may reinforce the awareness of income disparities, in the worst case creating aid dependency or patron–client relations that could contribute to a keener sense of difference rather than a stronger discourse of regional identification.

Despite the long history of Arctic trans-nationalism, the Arctic regimes examined in this chapter represent stable and permanent forums in which cooperation and a sense of regionality among indigenous leaders and others in their organizations can be developed, maintained and fine-tuned. Importantly, Arctic Council activities are tasks around which indigenous cooperation with other indigenous organizations and member states can be focused. In this way, sporadic or ad hoc cooperation can be reduced, leading towards a denser discursive interactiveness as indigenous organizations work to present a shared position.

Furthermore, as noted above, both the Arctic Council and the BEAR have provided reasons for bringing in and supporting new Arctic indigenous organizations – particularly in Russia. This can certainly be construed as an important step towards increasing the flow of people and ideas within the Arctic region and building up a bloc of organizations that can work together to promote indigenous interests within Arctic regimes and other arenas. However, such interactiveness between organizations should not be taken to indicate an increased sense of identification amongst the Arctic indigenous population as a whole. While some organizations may have quite well-developed communication with the indigenous peoples they are meant to represent, and may succeed in promoting a sense of Arctic regionality amongst them through the media or meetings that draw a wide audience, such a link should not be automatically assumed.

Conclusions

The Arctic Council and the BEAR have had little impact on rights to land and natural resources – perhaps the greatest immediate problem faced by indigenous peoples, at least in the Nordic countries and in Russia. In this sense, the regimes themselves are incomplete in relation to the problems at hand. On the other hand, it could be said that these regimes, particularly the Arctic Council, fill a particular niche in managing indigenous-government trans-national relations in the Arctic region and in dealing with trans-boundary problems that affect indigenous peoples. To take one example, indigenous organizations active in the Arctic Council are primarily mandated to deal with international issues. Movement of the ICC into specific questions of land use in Canada would probably result in resistance from other domestic Inuit organizations (who would consider such issues 'their issues'), as well as in significant and perhaps unproductive redundancy amongst various Inuit organizations. The ICC, RAIPON and the Saami Council do, however, deal with questions of indigenous land rights at the UN Working Group on Indigenous Populations. Perhaps the notion of niche applies here as well – recognizing that Arctic institutions are oriented toward issues related to sustainable development and regional cooperation, indigenous organizations deliberately tailor their efforts to maximize the potential benefits from these particular regimes.

The relative functional broadness of the Arctic Council and the BEAR, i.e. the wide range of issue areas addressed, also provides advantages in terms of linking indigenous issues to other issues. The Arctic Council, the BEAR and the matters discussed within these regimes, such as environmental problems, are seen to be high priorities by the member states. This is in contrast to the relative disinterest of Arctic states to committing themselves internationally on indigenous issues, as indicated by their non-ratification/non-implementation of ILO Convention 169. By including indigenous issues among the many other matters under their purview, the Arctic Council and the BEAR have created thematic bundles that, unlike ILO 169, are too big and important to be ignored by the states. Consequently, these regimes represent an opportunity for indigenous organizations to create explicit linkages between these broader issues like climate change, and indigenous issues.

In terms of representation, both the BEAR and the Arctic Council have been lauded as fundamentally innovative regimes because they have opened up to non-state actors areas that had previously been the exclusive preserve of foreign ministries. Interestingly, this reflects a trend in which the degree and quality of indigenous representation provided for in the Arctic regimes depends largely on when the regimes were established (Fjellheim 2003: 6). For example, there is no indigenous representation in the Nordic Council, whereas there is greater indigenous representation within the BEAR and rather significant representation of indigenous peoples in the Arctic Council. This indicates how Arctic regimes were influenced by the ever developing prominence of indigenous organizations and non-state actors in international institutions like the UN. Furthermore, this increasing representation over time perhaps indicates that the coexistence of partly overlapping regimes makes for healthy competition that can improve international standards.

When the BEAR was established it also made some sense, due to the emphasis on local actors, that the indigenous peoples in the region should be represented on the same level as the provincial authorities. However, the more state-focused structure of the Arctic Council provides for a higher and more effective level of indigenous representation. Although the indigenous actors cannot vote, they sit at the same table with the states and gain essential access to representatives of states. Furthermore, the fact that permanent participants within the Arctic Council have generally been able to participate fully in both Arctic Council negotiations and working groups means that indigenous organizations can exert influence on agenda-setting and on the design and implementation of Arctic Council programmes. Consequently, state representatives, in working to promote their own platforms and positions, have reason to build alliances with indigenous organizations. These features contribute to the effectiveness of the Arctic Council in addressing indigenous problems, primarily because indigenous organizations have the authority needed to represent themselves effectively, and to work for the development of new forms of political interaction in the Arctic. Effective indigenous participation also creates a situation in which it is beneficial for more advanced indigenous organizations to assist emerging ones. In the BEAR, by contrast, the motivation to build such state–indigenous partnerships and the opportunities for indigenous organizations to represent themselves effectively are largely absent. Thus, it seems that it is not only the fact of representation that is important in promoting effective responses to indigenous problems and in generating new forms of political mobilization, but its quality and degree as well.

It is important to note that there is also a potential downside to the closer contact and relationships between indigenous organizations and states catalysed by the Arctic Council. Indigenous organizations and states remain unequal partners, and closer relationships between them may also result in dependence, financial or otherwise, on the part of the indigenous organizations. A detailed examination of indigenous representation in the Arctic Council and the International Whaling Commission by Blichfeldt (2005: 120) shows how indigenous representation in such organizations can also be seen as a way for states to appear to be engaging cooperatively with indigenous organizations – without actually taking indigenous interests properly into account.

The research programmes carried out under the Arctic Council have contributed valuable and widely accepted scientific data from which indigenous organizations can develop policy and arguments. Case studies of ICC advocacy in relation to POPs and climate change clearly demonstrate the value of such Arctic-specific information in increasing the efficacy of the ICC in regional and international politics. Furthermore, such research programmes contribute to region building by validating the claims of indigenous organizations to have the Arctic recognized as a special region particularly deserving of international attention and concern. Accomplishing such comprehensive research programmes would be beyond the capacity of any one state, and certainly of any indigenous organization. Furthermore, it is unlikely that such research would have been carried out by a non-Arctic international regime. In this respect the Arctic Council has added something new and valuable to Arctic indigenous politics.

One of the BEAR's most prominent areas of action in relation to indigenous peoples has been a range of aid and development projects directed primarily towards the indigenous peoples of the European Russian North. It is relatively easy to take aim at the vast project portfolios of the Barents institutions, particularly as the problems facing Russia's indigenous peoples are not clearly trans-national in nature (like the issues of climate change or pollutants taken up in the Arctic

Council). The indigenous peoples in Russia are struggling with under-developed or weak relationships to various levels of government, and with unrecognized or poorly implemented rights to land and resources. Such issues cannot be addressed by improving the façades of buildings or by promoting 'aid tourism' that moves primarily from West to East. While these interactions contribute to the construction of discourses about regionality and certainly a denser interactiveness across the borders that divide the region, such projects cannot help to mitigate the problems faced by indigenous peoples of the region nor in improved political mobilization. In this sense, the 'fit' between the institution and the problems at hand is problematic.

The BEAR does fill a niche, however, as the project funding that it can provide is unlikely to be obtained elsewhere, in Russia or abroad. And while these projects may not have had a large impact individually, the cumulative sum of impacts may yet prove significant.

Notes

1 This chapter is primarily based on reviews of secondary and primary literature (mostly in the form of organizational documents), but also on extensive prior fieldwork and numerous interviews in the Barents Region and in the northern areas of Russia and Canada. Indra Øverland has worked extensively in the Barents Region, particularly among the Russian Saami on the Kola Peninsula in Russia but also in northern Norway and Finland. Elana Wilson has conducted field research in Nunavut (northern Canada) and Khanty-Mansiysk Autonomous Okrug (Russia) related to Canada-Russia bilateral Arctic cooperation and international development programmes.

2 This figure is from Russian Federation (2000). That document mentions a total of 45 numerically small indigenous peoples outside Dagestan, 39 of which are northern.

3 Interestingly, discussion of the Marine Mammal Protection Act (MMPA) seems to have remained off the Arctic Council agenda despite the inclusion of permanent participants. At the very least, no discussion of the MMPA is evident in the Council's published documents. This US legislation, enacted in 1972, is controversial in that it prohibits the import of marine mammals and any products derived from them into the USA, and represents a serious obstacle to the Inuit sealing industry in Greenland and Canada. ICC representatives have worked through a variety of channels to convince the USA to withdraw the ban entirely (Bourgeois 1998). However, these ICC efforts have not involved full-scale diplomatic or public support campaigns, because Alaskan members of the ICC argue that the MMPA is the only legislation that supports native Alaskans' rights to subsistence hunting (Bourgeois 1998; George 1999). Even should the ICC internal debate resolve itself, it is unlikely that the legislation will be discussed within the Council. The US government responded to the

harvesting of two whales by Inuit in Nunavut in 1994 and 1996 and to pressure from environmental and animal rights groups by vetoing any discussion of marine mammal management issues in the Arctic Council (ICC 2004).

4   Tennberg has argued that the Saami Council and RAIPON, two of the initial three indigenous organizations involved in the inception of the Arctic Council, have not been able to participate as effectively as the ICC. RAIPON struggled with changing internal goals and structure and lack of translated materials in the early days of the Arctic Council. A Saami representative noted that meaningful participation is curtailed by lack of resources and lack of specified personnel to represent Saami at the Council (their representatives, unlike ICC's designated personnel and political leaders, participate on a voluntary basis) (Tennberg 2000).

5   For the texts of these conventions see:
    –   LRTAP: http://www.unece.org/env/lrtap/full%20text/1979.CLRTAP.e.pdf
    –   Århus Protocol: http://www.unece.org/env/lrtap/full%20text/1998.POPs.e.pdf
    –   Stockholm Convention: http://www.pops.int/documents/convtext/convtext_en.pdf

6   Indigenous organizations in Canada were frustrated by what they saw as the failure of the federal government to fulfil its obligation, outlined in treaty agreements, to consult with indigenous peoples on international issues affecting their rights. After several ICC and coalition interventions, the coalition reached an understanding with the federal agencies involved and participated fully in the negotiations process established by UNEP that led to the signing of the Stockholm Convention (Fenge 2003; Watt-Cloutier 2003).

7   The bilateral character of most activities under the BEAR is particularly striking since one main reason for establishing the multilateral BEAR was precisely to avoid a bilateral framework (Lille 2004; Jervell 2002).

8   See http://www.arcticathabaskancouncil.com/intro/index.php

9   See http://www.gwichin.org/

10   See http://www.arctic-council.org/en/main/infopage/36/

11   See http://www.csipn.ru/about/en/inripp.htm

12   We are thus taking a slightly different line here from Young and Einarsson (2004: 17) in the *Arctic Human Development Report*, who do not mention the importance of contacts between indigenous peoples for the formation of an Arctic regional identity.

References

Arctic Council (1996) *Declaration on the Establishment of the Arctic Council*, Online. Available HTTP: <http://www.arctic-council.org/en/main/infopage/190/> (accessed 25 November 2005).

Arctic Council (1997) *The Alta Declaration*, Online. Available: <http://www.arcticcouncil.org/en/main/infopage/199/> (accessed 25 November 2005).

Arctic Council (2004) *Arctic Climate Impact Assessment Policy Document*, Online. Available HTTP: <http://www.amap.no/acia/index.html> (accessed 25 November 2005).

Blichfeldt, Georg (2005) 'På like fot? Om stater og urfolk, vitenskap og politikk med utgangspunkt i to case: Hvalfangstkommisjonen og Arktisk råd', unpublished Master's Thesis, University of Tromsø.

Bogoyavlenskiy, D. and Siggner, A. (2004) 'Arctic demography', in Niels Einarsson, Joan Nymand Larsen, Annika Nilsson and Oran Young (eds.), *Arctic Human Development Report*, Akureyri: Stefansson Arctic Institute.

Bourgeois, A. (1998) 'Alaskans disagree with Canada-Greenland position on MMPA', *Nunatsiaq News*, 30 July.

Downie, D.L and Fenge, T. (2003) 'Introduction', in D. Downie and T. Fenge (eds.), *Northern Lights against POPs: Combating Toxic Threats in the Arctic*, Montreal and Kingston: McGill–Queen's University Press.

European Environment Agency (2004) *Arctic Environment: European Perspectives. Why Should Europe Care?*, Copenhagen: European Environment Agency.

Fenge, T. (2003) 'POPs and Inuit: Influencing the global agenda', in D. Downie and T. Fenge (eds.), *Northern Lights against POPs: Combatting Toxic Threats in the Arctic*, Montreal and Kingston: McGill–Queen's University Press.

Fjellheim, R. S. (2003) *Urfolk i Arktis – Noen politiske utfordringer (Background note for the Committee on the North)*, Karasjok: Jaruma.

Foreign Affairs Committee (Norwegian Parliament) (2001) *Instilling fra utenrikskomiteen om nordisk samarbeid*, Report (S) No. 11, Oslo: Stortinget.

George, J. (1999) 'ICC split over Marine Mammal Protection Act', *Nunatsiaq News*, 30 November.

George, J. (2005) '"Climate warrior" plots strategy on behalf of Inuit', *Nunatsiaq News*, 18 November.

ICC (2004) 'Abstract Six: Marine Mammal Trade and a Conflict of Philosophies' Online. Available HTTP: <http://www.thectk.com/case/abstract06_04.html> (accessed 26 January 2006).

ILO (1989) *Convention 169 Concerning Indigenous and Tribal Peoples in Independent Countries*, Geneva: ILO.

IPS (2005) *Arctic Council Indigenous Peoples' Secretariat Homepage*, Online. Available HTTP: <http://www.arcticpeoplesorg> (accessed 20 November 2005).

Innukshuk, R. (1994) 'Inuit and self-determination: the role of the Inuit Circumpolar Conference', in W.J Assies and A.J. Hoekema (eds.), *Indigenous Peoples' Experiences with Self-Government*, IWGIA Document 76, Copenhagen: IWGIA and the University of Amsterdam.

IWGIA (2001) *The Indigenous World*, Copenhagen: IWGIA.

Jervell, S. (2002) '10 Years of the Barents Cooperation', in O. Pettersen (ed.), *The Vision that Became Reality: The Regional Barents Cooperation 1993–2003*, Kirkenes: Barents Secretariat.

Jull, P. (1999) 'Indigenous internationalism: what should we do next?', *Indigenous Affairs*, 1: 13–17.

Keskitalo, E.C.H. (2004) Negotiating the Arctic: *The Construction of an International Region*, London and New York: Routledge.

Landsdelsutvalget (2005) *For Nord! Ufordringer og muligheter for nordområdene med særlig vekt på Barentsregionen*, Bodø: Landsdelsutvalget.

Lille, C. (2004) *Samarbeid på viddene? En sammenlignende analyse of asatsningsområdene kultur, miljø og økonomi i Barentssamarbeidet*, unpublished Cand. Philol. thesis, University of Tromsø.

Maaka, R. and Fleras, A. (2000) 'Engaging with indigeneity: Tino Rangatiratanga in Aotearoa', in D. Ivison, P. Patton and W. Sanders (eds.), *Political Theory and the Rights of Indigenous Peoples*, Cambridge: Cambridge University Press.

MFA (2005) *Muligheter og utfordringer i Nord*, White Paper No. 30, Oslo: Royal Norwegian Ministry of Foreign Affairs.

Mitchell, M. (1996) *From Talking Chiefs to a Native Corporate Elite: The Birth of Class and Nationalism Among Canadian Inuit*, Montreal and Kingston: McGill–Queen's University Press.

Muehlebach, A. (2003) 'What self in self-determination?: Notes from the frontiers of transnational indigenous activism', *Identities: Global Studies in Culture and Power* 10: 241–268.

Murashko, O. (2002) 'Introduction', in T. Kohler and K. Wessendorf (eds.), *Towards a New Millennium: Ten years of the Indigenous Movement in Russia*, IWGIA Document 107. Copenhagen: IWGIA.

Nilsen, Thomas (2004) *Indigenous Peoples: Humiliating Financial Situation*, Barents Secretariat Press Release, 3 March.

Nordic Council (2005) *Nordic Council and Nordic Council of Ministers Homepage*, Online. Available HTTP: <www.norden.org/start/start.asp> (accessed 6 December 2005).

NOU (2003) *Mot Nord! Utfordringer og muligheter i nordområdene: Ekspertutvalg nedsatt av regjeringen 3. mars 2003*, No 32. Oslo.

Nunatsiaq News (1998) 'ICC pleads for reduction in transboundary poisons', 2 July.

Nuttall, M. (1998) *Protecting the Arctic: Indigenous Peoples and Cultural Survival*, Amsterdam: Harwood Academic.

Nystad, A. E. (2003) *Urfolksarbeidet i den Euro-Arktiske Barentsregion*, Kirkenes: Barents Secretariat.

Offerdal, Kristine (2003) 'Slutten for Barentssamarbeidet?', *Nordlys*, 19 December.

Prakhova, Anna, Alf Nystad and Roman Mikhalyuk (2004) *Preliminary Report of the Working Group of Indigenous Peoples in the BEAR (WGIP) 2002–2003: To Be Submitted to the Regional Committee*, Kirkenes: Barents Secretariat.

Russian Federation (2000) *O edinom perechne korennykh malochislennykh narodov Rossiiskoy Federatsii*, governmental decree, March 24.

Tennberg, M. (2000) *Arctic Environmental Cooperation: A Study in Governmentality*, Burlington, VT: Ashgate.

Toivanen, Reetta (2002) *Defining a People: How Do International Rights Influence the Identity Formation of Minority Groups?* Harvard University Center for European Studies Working Paper No 84.

Watt-Cloutier, S. (2003) 'The Inuit journey towards a POPs-free world', in D. Downie and T. Fenge (eds.), *Northern Lights against POPs: Combatting Toxic Threats in the Arctic*, Montreal and Kingston: McGill–Queen's University Press.

Watt-Cloutier, S. (2005a) 'Climate Change and the Arctic: Our Communities, Our Lives', Keynote Address to Snowchange 2005 in Anchorage Alaska, Online. Available HTTP: < http://www.inuitcircumpolar.com/index.php?ID=313&LANG=En> (accessed 25 November 2005).

Watt-Cloutier, S. (2005b) 'Remarks to the United Nations Environment Programme at the "Champions of the Earth" Award Ceremony, New York, April 19, 2005', Online. Available HTTP: <http://www.inuitcircumpolar.com/index.ph;?ID=294&Lang=En> (accessed 21 November 2005).

Watt-Cloutier, S. (2005c) *Petition to Inter American Commission on Human Rights Seeking Relief from Violations Resulting from Global Warming Caused by Acts and Omissions of the United States*, Online. Available HTTP: <http://www.earthjustice.org/news/documents/12–05/FINALPetitionICC.pdf> (accessed 12 December 2005).

Wiben Jensen, Marianne (2004) 'Editorial', *Indigenous Affairs* 4: 4–7.

Wilson, E. (2006) *Building an Arctic Community of Knowledge: The Promotion and Reception of Canadian Resource Management and Economic Development Models in the Russian North*, unpublished D.Phil Thesis, University of Cambridge.

Young, O. (1992) *Arctic Politics: Conflict and Cooperation in the Circumpolar North*, Hanover, NH and London: University Press of New England.

Young, Oran and Einarsson, Niels (2004) 'Introduction', in Niels Einarsson, Joan Nymand Larsen, Annika Nilsson and Oran Young (eds.), *Arctic Human Development Report*, Akureyri: Stefansson Arctic Institute.

4 Communicable disease control

Lars Rowe and Geir Hønneland

Introduction

The bloodless and hopeful revolution that culminated in the collapse of the Soviet Union was accompanied by a looming disaster. In the wake of enthusiastic reforms and shock therapy, the transitional challenges that confronted post-Soviet societies became increasingly apparent. The fallacies and the structural legacy of the socialist mega-state weighed heavily on the shoulders of its heirs, and could not be rectified speedily. In the Soviet planned economy, commodities had been bought and services rendered – albeit inadequately – under the protective wings of the state. The Soviet social contract was built on the basic premise of a paternalistic state that took care of its underlings. The healthcare system, named after and built on the ideas of the first Soviet health commissar N. A. Semashko, reflected this premise: In principle, and to a large extent in reality, all Soviet citizens were provided with free access to health care in their local community (Twigg 2000: 43). With the abrupt and unplanned disintegration of the Semashko system and the evaporation of necessary funding from the central Ministry of Health, health care was cut to a minimum overnight. The consequences were alarming indeed.

In this chapter, we describe the deterioration in public health that has become a prominent feature of post-Soviet societies, and how the problems arising from the collapse of healthcare systems in Northwest Russia and the Baltic states have been countered under the auspices of two Arctic regimes. Both the BEAR and the CBSS have established programmes aimed at evening out the differences in health standards across the shared borders of Northeast Europe. A more recent development which will also be examined is the attempt to establish a Partnership in Public Health and Social Wellbeing as part of the European Union's Northern Dimension (EUND). Also, the Human Health Assessment

Group, a sub-programme under the Arctic Council's Arctic Monitoring and Assessment Programme (AMAP), has been involved in health-related work. The activities of this group, however, are more directed towards monitoring and assessment of environmental contaminants, not actual on-location health projects. The group is therefore defined in this book not as a health initiative but rather as an environmental programme.

The emphasis in this chapter is on the countermeasures taken specifically against the spread of communicable diseases, such as HIV and tuberculosis, as the above-mentioned initiatives have focused on this particular area. However, the recent tendency to broaden the scope of such health-related programmes will also be taken into account. We then go on to comment on the effectiveness of the various programmes – without venturing to gauge their impact on human health in the region, as that would transgress both the limited scope of this chapter as well as the studies we have conducted in the field of public health programmes in the Arctic. We will, through case studies, comment on the mobilization of new groups that have been given the opportunity to interact with colleagues in the region through collaborative health efforts. Finally, we comment on the contribution made to region building through these efforts.

This chapter presents selected findings from several research projects. In 2002, Geir Hønneland and Arild Moe produced the *Evaluation of the Barents Health Programme* (Hønneland and Moe 2002). From 2000 to 2004, Geir Hønneland and Lars Rowe conducted a 'contextual evaluation' of the Task Force on Communicable Disease Control in the Baltic Sea Region, published in book form as *Health as International Politics* (Hønneland and Rowe 2004). A follow-up study was conducted by Lars Rowe and Bernd Rechel in 2005, resulting in the article 'Fighting Tuberculosis and HIV/AIDS in Northeast Europe' (Rowe and Rechel 2006).

Our studies have been based on a qualitative approach, using two complementary research methods: a documentary analysis and a large number of semi-structured in-depth interviews with key stakeholders, to permit data triangulation. From 2002 to 2004 we conducted approximately 100 interviews with deliberately selected Task Force participants during field visits throughout Northeast Europe. The interviewing was carried out by Rowe and Hønneland. In the follow-up study in the summer of 2005, a series of interviews with key actors in the then-terminated Task Force collaboration added to our knowledge base.[1]

The problems faced

By the late 1990s, the health situation in Russia and the Baltic states was causing serious concern among medical experts and officials in the West.[2] Life expectancy had decreased dramatically since the break-up of the Soviet Union, mainly as a result of diseases caused by malnutrition, smoking and/or alcohol consumption.[3] In addition, tuberculosis, which in Western societies had been more or less eliminated or at least controlled effectively, was re-emerging, and HIV/AIDS was already causing widespread suffering in the post-Soviet area. Centuries-old fears of infectious diseases spreading like wildfire from person to person and country to country were being rekindled. Although Russia and the Baltic states were considered to be most at risk from tuberculosis and HIV/AIDS, some commentators went so far as to suggest that the epidemics could destabilize the political climate in Northern Europe as a whole.

The most severe effects of both epidemics were felt in Russia. For tuberculosis, after levelling out at around 30 cases per 100,000 population in the early 1990s, the rate rose dramatically towards the end of the decade, reaching approximately 80 reported cases per 100,000. The HIV figures were even worse, rising steeply towards the end of the 1990s, and prompting dramatic statements from Western medical experts and news agencies. While reported cases per million population had been just below 25 in mid-1997, a disturbing 130 cases were recorded by the end of 1999, and a doubling by the year 2000 was indicated. In the Baltic states, the situation was less dramatic, but still serious. In Estonia, while the number of tuberculosis cases was relatively low, the country suffered an outbreak of HIV among injecting drug users in 2000. Latvia paralleled the Russian case regarding both tuberculosis and HIV. In Lithuania, although tuberculosis rates reached Russian and Latvian levels, the HIV rate remained as low as that of Sweden across the Baltic Sea.

The entire situation was deemed unacceptable by Western experts, particularly the emergence of a multi-drug resistant tuberculosis strain, caused by insufficient or interrupted treatment of ordinary tuberculosis. Terms like 'worst-case scenario' and 'unprecedented public health risk' were used, as were terms like 'katastroika' and 'mortality crisis'. The worried Western community emphasized the correspondence between economic recession and high tuberculosis and HIV/AIDS incidence in the former Soviet areas. It was feared that the social disruption could threaten the eastern part of the Baltic Sea region and might destabilize the area not only socially, but politically as well. The health crisis was expected to damage interstate relations in the eastern areas of the region. Without basic security in terms of human health, it was claimed, basic security in the wider social and political sense could not be achieved.

Regime-based responses

The traditional method of preventing the cross-border spread of infectious diseases has been to enforce a strict quarantine regime combined with equally strict medical screening at national borders. But as nation states have become increasingly reliant on commercial and cultural interaction, such methods have lost much of their appeal. More specifically, the post-Cold War ambition of encouraging rather than discouraging cross-border contact in the European North rendered border-control measures both impractical and ideologically unacceptable. Thus, the path chosen in the fight against communicable diseases in the region would be through more, as opposed to less, human contact and cooperation.

The Barents Health Programme⁴

The Barents Euro-Arctic Council meeting in Luleå, Sweden, in January 1998 stated in its communiqué: 'several national Governments as well [as] the Regional Council have decided to give priority to health issues. Special attention should be paid to joint actions that will lead to rapid improvements in the health situation' (Barents Euro-Arctic Council 1999: 10). Accordingly, the Health Co-operation Programme in the Barents Euro-Arctic Region 1999–2002 (hereafter the Barents Health Programme) was established. The programme did not create new multilateral structures; it is based on bilateral projects and projects carried out by international organizations. Project coordination was to be carried out with the help of an international reference group through the exchange of information facilitated by the database Barents Information Service, administered by the Barents Secretariat in Kirkenes.

The Barents Health Programme has established a rather long list of objectives, activity areas, main guidelines and general project criteria, as well as specific project criteria and sub-goals (Barents Euro-Arctic Council 1999). On the basis of a general picture of the health-related situation in Northwestern Russia, five fields of activity were singled out in the programme:

1   combating new and re-emerging infectious diseases;
2   supporting reproductive health care and child health care;
3   counteracting lifestyle-related health problems;
4   improving services for indigenous peoples;
5   quality improvement of medical services.

Three main guidelines or principles were also established:

• Special attention should be paid to joint actions that will lead to rapid improvements in the health situation.
• Within all priority areas, special attention should be given to projects focusing on children.
• The health programme must support existing and future bilateral and multilateral health projects under the umbrella of the Barents Euro-Arctic Council.

These guidelines are formulated in a general way and must be understood as criteria meant to steer the selection of projects in all areas of activity.

Norway allocated a total of EUR 5.9 million to projects in the Barents Health Programme (Hønneland and Moe 2002: 3). The following discussion of distribution among programme areas refers to this sum, which has been published (as opposed to the contributions of the other countries) and is presumably substantially larger than the allocations of other Barents member states. The programme did not specify any particular distribution of resources among the activity areas, although it would have been reasonable to expect substantial efforts in all five areas. However, as depicted in Figure 4.1, this was not to be the case. The first two areas have predominated, with 'Combating new and re-emerging infectious diseases', and 'Supporting reproductive health care and child health care' receiving 39 and 36 per cent of total funds respectively. Area 5 'Quality improvement of medical services' has received considerably less – 22 per cent. The striking feature is that the two remaining areas – area 4 'Improving services for indigenous people' and area 3 'Counteracting lifestyle-related health problems', received miniscule funding – 2 and 1 per cent respectively.

This disparity may have several explanations. Characteristics of the first two activity areas as opposed to the three others may be important. Areas 3 and 4 may be harder to reconcile with the priority given to 'joint actions that will lead to rapid improvements in the health situation'. Especially 'counteracting lifestyle-related health problems' seems to necessitate a long-term effort. That said, the same could also

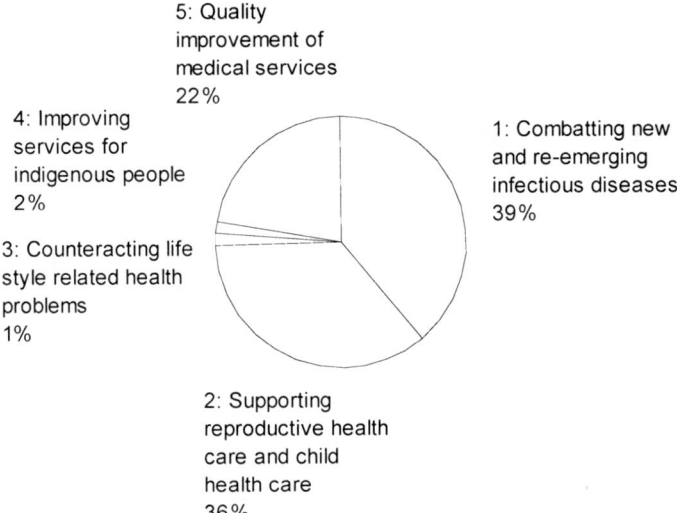

5: Quality
improvement of
medical services
22%

4: Improving
services for
indigenous people
2%

3: Counteracting life
style related health
problems
1%

1: Combatting new
and re-emerging
infectious diseases
39%

2: Supporting
reproductive health
care and child
health care
36%

Figure 4.1 Budget allocations under the Barents Health Programme, 1999–2002.

be said of communicable diseases or reproductive health. In general, international health-related aid programmes have tended to overemphasize communicable diseases, at the expense of non-communicable diseases (Suhrcke *et al.* 2005: 925). This is also reflected in the UN Millennium Development Goals pertaining to public health, where reproductive health and communicable diseases are strongly in focus, whereas non-communicable diseases are disregarded (Rechel *et al.* 2004). Area 4 'Improving services for indigenous people' also seems under-prioritized. On the other hand, there is no denying that indigenous people represent a very small proportion of the total population, thereby perhaps warranting a smaller share of total funding.

The Barents Health Programme was extended for a second period, from 2004 to 2007. Some adjustments were made to the priorities in the new programme:

- prevention and combat of communicable diseases;
- prevention of lifestyle-related health and social problems and promotion of healthy lifestyles;
- development and integration of primary health care.

The emphasis on countering the spread of communicable diseases remains strong. The prime ministers in the Barents region made special mention of the fight against tuberculosis in a statement from the 10th anniversary of the BEAR, in which they appealed to the health workers of the region to strive towards complete control over the disease within the next ten years. Lifestyle-related health was also a part of the first Barents Health Programme, but, as we noted, was allocated no more than 1 per cent of the total budget. That it is now brought up again, may signal greater efforts in this field. The third point, 'development and integration of primary health care' is new and clearly reflects the realities of Russian health services. In line with the general organization of Soviet society, Russian health services have traditionally been 'compartmentalized' and not vertically integrated between the different sectors of public life. This means that health services established within, e.g., the military sector or the prison sector did not cooperate with the civilian sectors, nor did medical personnel from the two sectors communicate. This has now become a problem, as patients move freely from one sector to the other (Hønneland 2005: 95). Integrational efforts were also part of the Task Force on Communicable Disease Control (see below) which most likely has been 'setting a trend' for the new Barents Health Programme.

Task Force on Communicable Disease Control in the Baltic Sea Region[5]

The Task Force was established in April 2000 at the third summit of CBSS in Kolding, Denmark.[6] It was meant to be an intensive short-term effort, responding to the widely acknowledged threat of communicable diseases in the Baltic Sea region. The choice of organizational model was motivated by the desire to avoid yet another cumbersome interstate bureaucracy. The basic idea behind the Task Force organization was to secure high-level political support from all participating countries, without losing sight of the grassroots level. While a hierarchical structure was set up, headed by appointed representatives of the 11 member governments and the European Commission, main activity was concentrated on the ground, where medical experts at the local, regional and national level were given ample opportunities to develop and implement projects. To ensure communication between the operational and the political level, a group of senior health officials was established as an intermediate body between the highest political level and the six topical programme groups.

The programme groups, consisting of medical experts of participating states, were the operational core of the programme. They covered the main areas addressed by the Task Force:

1 epidemiological surveillance
2 tuberculosis
3 HIV/AIDS and Sexually Transmitted Diseases
4 antibiotic resistance and hospital infection control
5 primary health care
6 and (established later), prison health

(Task Force 2004)

The programme groups were meant to be the link between projects and funding sources. Their members were responsible for approving projects developed by individual group members or local medical workers. The programme groups constituted, by their very existence, one of the most important achievements of the Task Force – the establishment of international medical networks in Northeast Europe. We return to this in the section on mobilization.

The Task Force managed to initiate and implement a substantial number of projects. Table 4.1 shows the number of projects across programme areas and countries at the end of 2003.

Although the Task Force did continue in 2004, not many new projects were initiated. About half of the 213 projects were carried out in Russia, reflecting its geographical size and large population, as well as the crisis experienced by the country's healthcare system.

A major health collaboration like the Task Force requires both political commitment and financial support from participating states. In the founding document, the Norwegian initiators envisaged the allocation of EUR 25 million for the first year of a proposed three-year period (Task Force 2000: 42). In its final report, however, the Task Force noted that only a total of EUR 18–20 million had been secured over the four-year period, a figure including both administrative expenses and allocations to projects, the latter amounting to EUR 9–10 million (Task Force 2004: 8). This equals an annual average of EUR 5 million per year, so the actual willingness of participating Western countries to allocate resources to the Task Force has fallen far short of the aspirations of the programme's initiators. The mandate of the Task Force officially ended in June 2004, with an expressed desire to continue along the lines pursued thus far.

The Northern Dimension Partnership in Public Health and Social Wellbeing[7]

With the aim of continuing the Task Force legacy and expanding the health collaboration to include non-communicable diseases, the

Table 4.1 Overview of projects under the Task Force 2000–03

Host Country	Number of projects [a] Surveillance	Tuberculosis	HIV/STD	Antimicrobial Resistance	Primary Health Care	Prison Health	Total
Estonia	1	5	11	6	9	4	36
Latvia	2	5	6	8	11	3	35
Lithuania	2	3	13	7	13	7	45
Poland	0	0	4	1	9	4	17
Russia	19	17	28	9	19	13	105
Trans-national	5	1	2	1	0	1	10
Not specified	0	0	1	3	4	1	9
Sum [b]	29	31	65	35	65	33	257
Projects	25	31	49	31	46	31	213

Notes:
[a] The table contains projects that were 'closed for other reasons', i.e. rejected or cancelled after initial approval.
[b] The sum exceeds the number of projects within a programme area because some projects are implemented in more than one country.

Northern Dimension Partnership in Health and Social Wellbeing (hereafter 'the Partnership') was created in October 2003. The Partnership added France and Canada to the 11 Task Force countries and included eight partner organizations.[8] Some of these organizations, which included what are in this book referred to as 'Arctic regimes' like the CBSS and the Barents Euro-Arctic Region, were already involved in health programmes. Thus, the Partnership was designed to be a project-oriented body, but perhaps the most important intention was to create a coordinating body.

On 17 October 2003, the European Council endorsed the Second Northern Dimension Action Plan for 2004–06,[9] aimed at addressing the special regional development challenges of Northern Europe, intensifying cross-border cooperation between the EU and its neighbours and establishing closer regional-level relations with Russia. The Plan focuses on five sectors, of which one includes health issues: 'human resources, education, culture, scientific research and health'.[10] Here the fight against HIV/AIDS and tuberculosis is mentioned as the most outstanding health issue.[11] The Declaration concerning the Establishment of a Northern Dimension in Public Health and Social Wellbeing was signed in Oslo on 27 October 2003.[12]

The Partnership set up an organizational structure paralleling the Task Force model, with a Partnership Annual Conference (PAC) as its highest body.[13] As a link between the expert groups (continuation of the Task Force programme groups) and the PAC, a Committee of Senior Representatives was established. The overall structure of the Partnership is inspired by and resembles the Task Force. There are two main differences: the intended mandate period, and the geographical scope. Whereas the Task Force was a time-limited effort, the Partnership is meant to serve as a permanent coordinating body for all activities in what is defined as the 'Northern Dimension' – which includes both the Barents and the Baltic Sea regions.

Effectiveness

As seen above, the *output* of Arctic regimes' efforts within the health sector has amounted to several organizational structures. However, measuring the *impact* of health collaborations is very difficult, if not impossible. A wide range of factors not connected with health collaborations can cause both degradation of and improvement in the public health situation (see Rechel and McKee 2004: 13–19). However, as an identifiable part of the *outcome*, behavioural adaptation, as mentioned in Chapter 2 in the present volume, may be a good indicator of the success of Arctic regimes in their efforts to promote better public health. Russia plays a pivotal role with regard to health programmes in the Arctic. It is, as noted above, the state where the post-Soviet decay in public health services has been most dramatic, so all programmes give high priority to the Russian health sector. In turn, this has meant that Russia as a country and individual Russian bureaucrats and health workers are now confronted with constant pressure from their Western counterparts to acquiesce to public health schemes of Western design.

A case study that may illustrate the behavioural adaptation that can result from such focused regional interaction concerns the reception of the WHO tuberculosis strategy Directly Observed Treatment with Short-course therapy (DOTS). This strategy was promoted in all tuberculosis projects run by the Barents Health Programme and the Task Force, and will guide tuberculosis projects in the Partnership. Overall, DOTS-guided tuberculosis projects have received more funding in these regional health programmes than any other focus area. As a treatment scheme developed within the framework of the WHO, DOTS should not be labelled 'Western' in the general sense of the word. After all, Russia is also part of the WHO, and has as such participated in promoting the strategy in developing countries. However, when DOTS has been promoted in Russia under health programmes sponsored by Western states, it has become an issue of contention between Russian and Western tuberculosis specialists, so we will treat it as a 'Western' element in discussing efforts to counter the spread of tuberculosis in post-Soviet countries.

Acceptance of DOTS in Russia is a good yardstick of the effectiveness of the Arctic health regimes. Of greater significance, of course, would be a quantitative improvement in tuberculosis and HIV/AIDS cases, but as mentioned, attributing to Arctic programmes any changes in the general health situation would be highly speculative, due to the complexities of causality. The DOTS strategy was the defining element in all tuberculosis projects, as well as being the single most contended issue between Russian and non-Russian tuberculosis specialists. As we will see below, implementing DOTS in Russia is not merely a matter of small adjustments to the existing system, but involves a core restructuring of the diagnostic principles and treatment methods applied. An established Russian medical tradition has been challenged – by outsiders – and the ensuing difficulties have been substantial. Thus, the degree of DOTS acceptance in Russia is not only a measure of treatment efficiency, but even more a measure of the ability of Arctic health programmes to break through behavioural patterns that have become entrenched through a long tradition of Russian tuberculosis treatment.

The DOTS strategy is in conflict with traditional Soviet tuberculosis surveillance and treatment, which was for a long time a highly successful strategy towards controlling the tuberculosis bacteria.[14] The main difference between the traditional Russian approach to tuberculosis detection and treatment and DOTS is that the Russian approach pursues active case-finding through mass screening of the population (instead of passive case-finding based on sputum smear microscopy among self-reporting patients); hospitalization and isolation (instead of outpatient treatment); long-term, individualized multi-drug approaches (instead of short-course standard cure); and the use of surgery no longer employed in the West. According to Russian thinking, diagnosis based on bacteriology is a diagnosis made too late, so radiography is the main method of diagnosis (Kimerling 2000: S162). According to Western experts, without improvements in case detection and cure, the incidence of tuberculosis in Russia is expected to rise by about 10 per cent per year, i.e. a doubling every seven years. The introduction and expansion of DOTS is expected to halt the rise, and by 2015 a 10 per cent annual decline in incidence is predicted, if DOTS is widely implemented (Coker 2001: 435).

Despite well-documented successes of DOTS implementation in developing countries, and the obvious need for new and cost-efficient solutions in Russian tuberculosis treatment, the barriers have been many. To start out with the general picture: DOTS has not been adopted as a general policy towards tuberculosis control in Russia. At the same time: DOTS programmes have been implemented in several Russian regions, and official Russian tuberculosis policy after the turn of the millennium is broadly similar to DOTS with its various components. In the Baltic states, DOTS has been integrated in national tuberculosis programmes.

Russian opposition to DOTS

One of the leading Russian experts on tuberculosis control, Mikhail Perelman, articulates Russian opposition to DOTS in an article from 2000 (Perelman 2000). His argument is largely twofold: first, he claims that certain aspects of the DOTS regime are not applicable in the Russian setting; second, he opposes the use of the acronym DOTS in general, and to signify something different from traditional Russian tuberculosis control in particular. The thrust of his argument is that hospitalization cannot be fully substituted by ambulatory treatment because patients with chronic or drug-resistant tuberculosis need to be treated in hospitals, and because in many parts of Russia that is more cost-effective than ambulatory treatment. The same argument applies to tuberculosis patients with subsidiary conditions like alcoholism and 'anti-social behaviour'. He also points to the disagreement between DOTS and the traditional Russian method concerning laboratory diagnosis, and the Russian insistence on surgery as part of tuberculosis treatment. Concerning the latter, he states: 'We anticipate an increase in the number of surgical interventions to achieve higher cure rates among the increasing numbers of chronic and drug-resistant cases' (ibid.: 1101).

Perelman's objection to the acronym DOTS goes as follows:

[F]rom our perspective, the western acronym DOTS merits some special comment. Our colleagues from the WHO and from other international organizations have attempted to integrate DOTS into Russian phthisiatry. Direct translation of DOTS (directly observed therapy, short course) into Russian is 'treatment with short course under direct observation', or 'controlled treatment with short course'. For the following reasons the DOTS acronym is considered unacceptable by many Russian phthisiopulmonologists, as well as by the author of this article.

1.   DOTS correctly reflects the meaning of only one of four principles of the antimicrobial therapy of tuberculosis, namely that it has to be controlled to ensure consistent drug administration.
2.   Two other important principles are not reflected in this acronym at all: one is the combined use of several drugs, another is the two phases of therapy, intensive and continuation.
3.   The emphasis on 'short course' is misleading. As opposed to other infectious diseases, the treatment of tuberculosis, in order to achieve good results, needs to be very long. The idea about fast cure through 'short course' chemotherapy (usually understood as a few days or weeks) is misleading and counterproductive, especially when taking into account the psychology of tuberculosis patients. The concept of directly observed therapy is well known to Russian physicians, and this principle has been implemented into routine practice for a long time. Therefore, the 'new' western acronym for this well-known principle of therapy has not been well received.

(ibid: 1102).

The tone in which the article is couched expresses a sense of indignation on behalf of Russia and traditional Russian (or Soviet) medicine. More than anything else, it seems that Perelman wants to show that WHO has not re-invented the wheel with its DOTS strategy. This sentiment is reflected, for instance, in his reference to 'the "new" Western acronym for this well-known principle of therapy'. Throughout the discussion, he also refers to DOTS as explicitly 'Western' (not merely 'foreign'), and to the ignorance of 'Western' experts of traditional Russian tuberculosis control.

Distrust of 'magic formulas' from the West is widespread in post-Soviet Russia. From the point of view of Russian tuberculosis experts, Russia has a proud heritage of tuberculosis control. This system has experienced problems since the break-up of the Soviet Union, but Russia is no developing country starting from scratch in its battle with the disease. Both the fact that DOTS is generally used in developing countries and the simplicity with which it is often presented by the WHO seem to exasperate Russian experts. In the West it is often a point in itself to ensure that information is provided as simply and accessibly as possible. The risk here is that Russian experts may feel they are being treated like amateurs, so they respond by looking for inconsistencies in the Western approach to flag their credentials. Hence Perelman's insistence on the DOTS acronym not covering all therapeutic aspects, and that a six-month course of treatment is 'not a short course, but a long course'. In sum, Perelman's scepticism to DOTS is not only an expression of patriotism or anti-Western sentiment ('we oppose DOTS because it comes from the West'), but a combination of national (or possibly personal) pride and a reaction to the way in which the alternative to the traditional Russian approach has been presented as a completely different approach which cannot be implemented only partly or adjusted to local requirements. From the WHO perspective, it is either all DOTS or no DOTS at all.[15]

The regions – Russian acceptance of DOTS

A conspicuous feature regarding Russia's introduction of DOTS is that the initiative has almost without exception come from regional authorities collaborating with Western organizations, such as the Barents Health Programme and Task Force (guided and condoned by the WHO) and not from federal Russian authorities.[16] The Arctic regimes are not the only players to introduce DOTS to Russia and the Baltic states, but the tuberculosis projects they have implemented have generally been oriented towards promoting DOTS. A noticeable feature is that the Baltic states and the Russian rim regions in the Northwest have embraced the WHO strategy wholeheartedly, while central tuberculosis experts in Moscow and St Petersburg display opposition in line with that of Perelman (2000). On the question of DOTS in Russia, a leading Russian tuberculosis experts in the Task Force collaboration, located in one of the 'capitals', stated:

> Ah, this is where [my Western collaboration partners] call me an old Communist. You know, DOTS consists of five elements. [. . .] We're not in any way against any of them. No sensible person would be. And we're grateful to the West for bringing it here. As a result, tuberculosis has become an important issue in Russia. But we do have a tradition in Russia, and not everything about it is bad!

She went on to speak about how unfair it is that regions adopting the DOTS strategy suddenly find themselves in a much more favourable situation financially than regions that retain the traditional Russian method: 'Once you declare your intention to introduce DOTS, your laboratory is redone up to Western standards and you receive a lot of financial support. It is rather unfair. Tuberculosis is a problem elsewhere, too.' She also emphasized that DOTS is not the real solution to the tuberculosis problem:

> Tuberculosis is a social disease. If we solve the social problems, we also solve the problem of tuberculosis. We have to think about why people don't want to recover. This is a problem that is not solved by DOTS. DOTS isn't adapted to the Russian reality. Look at other countries, like Sweden, Finland and Norway. They have solved the social problems and they don't have tuberculosis, but they never used DOTS.

Very much like Perelman, this expert seems primarily to dislike the way DOTS is profiled as some sort of 'magic formula', with the implicit assumption that none of the good things about DOTS can be found in the traditional Russian approach. She underscores the 'logical' basis of her scepticism and distances herself from those who reject DOTS on more nationalistic grounds: 'Many Russian doctors reject DOTS without really knowing what it is. [That's not right, but] the West should have presented the same contents without calling it DOTS.'

Neither the Baltic states nor northwestern rim of Russia are as dismissive of DOTS. The general picture drawn by our interviewees in these regions is encapsulated in the following paraphrased statement:

'The Finns [or Norwegians, or Swedes, or whatever] arrived, convinced us, and everything is going so much better now.' Representatives of the Russian regions that have introduced DOTS complain not about the West, but about the Russian tuberculosis establishment: 'In Russia, [DOTS] came from the regions (*s regionov shlo*). The problem lies with our Russian experts. They have to rid themselves of certain prejudices. The stereotypes have to be changed.' Another regional tuberculosis expert expressed moral indignation at the national establishment: 'As doctors, we are supposed to help people where we can. We promised that when we took the [Hippocratic] oath'. What she seems to be saying is that those who dismiss a proven strategy such as DOTS for unscientific, for instance patriotic reasons, are behaving immorally.

So was DOTS introduced in a number of Northwest Russian federal subjects simply because people from the West succeeded in convincing the Russians about the excellence of this strategy? There are other elements to the story. The relevance of the centre–region conflict is obvious: the DOTS issue is used by representatives of the Russian regions to oppose the superiority of the centre, understood as the federal capital of Moscow and the 'northwestern capital' of St Petersburg. The financial aspect, in turn an important component of centre–region relations, might also be more significant than immediately apparent. As expressed by one of our informants who had some involvement with the Task Force, although not with the actual DOTS-related projects:

> Before, we received strict orders (*prikazy*) from the centre. The orders were always followed by money, and we always complied with the orders. Suddenly, around 1998, there were the same federal orders, but no money. What were we to do? Then foreign organizations appeared with their DOTS projects. We decided to give it a try.

There is little doubt that both the Barents Health Programme and the Task Force have been instrumental in introducing the DOTS strategy in the Russian Arctic. If not always for purely medical reasons, but helped by factors like the ongoing struggle in Russia between the centre and the periphery and the financial benefits following the WHO strategy, the Arctic regimes can attribute the move towards DOTS in oblasts like Murmansk and Arkhangelsk, as well as in the Baltic countries,[17] to their activity. This constitutes a behavioural adaptation in the post-Soviet area that must be assumed to have a positive, albeit unquantifiable, effect on the health situation in the area.

Networks: Mobilization and region building

Close contact between medical personnel from various parts of the Arctic region was always thought to be a necessary step if other aims were to succeed in fighting communicable diseases. Not only would this ensure that local opinions and factors would be taken into account during project planning and development, it would also facilitate and possibly encourage the broad exchange of ideas and expertise among different medical communities with different traditions. In this section, we will discuss the mobilizing effect of the networks that were created in the health collaborations. The same networks, once they had mobilized actors in the healthcare sector, also helped to enhance region building in the Arctic.

The mobilization of medical and bureaucratic actors to interact across the borders, but also *within* the borders of some of the involved countries, has been held up as the most important achievement of the collaborative efforts in health. Horizontal integration between different sectors in the public health system in Russia has been strengthened (and to some extent established) as a consequence of health collaborations under the auspices of Arctic regimes. The encouragement of stronger interaction between the civilian health care and prison authorities has been central in both the Barents Health Programme and the Task Force (Hønneland and Moe 2002; Hønneland and Rowe 2004). Furthermore, the overriding perspective of primary health care attributed to all Task Force Projects through the Primary Health Care working group can be seen as an important step towards vertical, cross-sector integration in Russian health services.

Although the health collaborations were established to assuage concerns for public health in Northwestern Russia and the Baltic states, this was far from the only motivation. The official goal was to provide a public good – communicable disease control – but it embraced several other objectives as well. One was to help to 'construct' the Baltic Sea and Barents regions, that is, pave the way for more regional cooperation. But it also represented an attempt to attain certain objectives of Norway's foreign policy. As a non-EU member, Norway is constantly on the lookout for alternative multilateral arenas, to avoid being isolated internationally. Region-building efforts in Northern Europe after the Cold War (the BEAR and the CBSS) fit perfectly with this position.

There are, however, limitations to networking and integration. As we will see below, the sustainability of contacts that were established in the lifetime of a programme becomes highly questionable once the programme in question is terminated. For instance, the Partnership in Public Health and Social Wellbeing has not been able to sustain the networks it inherited from the Task Force. Furthermore, we find groups that could have been mobilized in the programmes, but were not. For example, the Barents Health Programme chose to emphasize the support of existing bilateral health projects. One consequence of this was that the programme would be less open to applications in areas with no or little cooperation in place, than in geographical or thematic areas that had existing partnership patterns to build on (Hønneland and Moe 2002).

As networking was most clearly emphasized as a precondition for the success of Task Force, we will use this as our case in the following. Networks can be more of a burden than an asset if they do not work well. In the case of the Task Force, the work devoted to widening cross-regional contacts seems to have been a wise choice. The networks put in place under the auspices of the Task Force were not only an immediate asset to the Task Force itself, they also represent a potential investment, and a self-sustaining undertaking after the collaboration was terminated. The programme groups are good illustrations of the networks developed under Task Force auspices. Although several of them report having experienced problems, these have mainly been of a professional nature. A central aspect of networks is that they permit personal relations to emerge and develop. This makes future contact not only more likely, but, according to many of our interviewees, even easy and pleasurable. One Latvian official ranked the networking endeavour as the most important achievement of the Task Force:

> The networks of the Task Force are important. That is what *really* works. [. . .] Many of the participants will continue collaborating in some form or other. We should keep the people together, but we need a new umbrella under which to work, a new international body, recognized by donors. If the Task Force networks are discontinued [. . .], it would be a disaster! – a lost opportunity! This is a true network, where trustful relations have been established, which is the most important thing. It will be very hard to create such a network once more – we should maintain this one.

Mobilization

The Task Force was successful in promoting enhanced interaction between medical workers and public health officials *within* the eastern parts of the Baltic region. At the national and regional levels and between Russia and the Baltic countries, better communication among medical workers is said to be a direct consequence of Task Force efforts. A Lithuanian official expressed satisfaction with the fact that Lithuanian medical workers were now working together, instead of keeping to their own district or municipality: 'Because of this, we now have a firm base of competent medical workers in Lithuania. Many got acquainted with the LFA [the Logical Framework Approach, a project design system applied in all Task Force projects], which is good, and they have met people working in the same way, within Lithuania and in other countries. Good contacts have been established.' In Russia, contacts between different oblasts have developed. Murmansk and Arkhangelsk Oblasts were successfully encouraged to work closer together, much to the satisfaction of the local Task Force officials.[18] This was, according to a Murmansk official, a radical departure from earlier times: 'It is not like before, when we went to Moscow and heard that other oblast representatives had been there [a few days or months ago]. Now we meet our regional colleagues face to face, at seminars and so on. And anyway, our meetings are now a lot less formal.' The interviewee was in no doubt that he preferred Western-style informality to Soviet-style formality. Certainly, the Soviet collapse has made it far easier for Russian specialists to meet their colleagues, with or without the foreign help. The federal authorities do not enforce the same form of discipline as their Soviet predecessors. That said, the Task Force did provide financial support to arrange seminars and meetings, precisely because it deemed personal contacts (the 'people-to-people' dimension) pivotal in the fight against communicable diseases.

Why, then, are the integrating networks here seen as a mobilizing factor? The answer is simple: Russian regional medical workers, the Baltic healthcare community and regional medical officials were all included in a programme that focused on their expressed needs. For instance, in Russian regions the requirements of the local tuberculosis dispensary or the regional AIDS centre would not be of particular importance in setting the allocated funding from central Russian authorities. Now, the health programmes not only enabled direct communication with a funding partner, they also provided a network through which wishes could be defined and developed within a set framework. The new networks could provide an arena for colleagues from different localities, regions and countries, and thereby have an empowering effect. This was particularly the case with Russian groups which had previously been merely a part of an oversized and inefficient healthcare system. Now they were given the tool to develop their own work place, more independently of their superiors in Moscow and St Petersburg. A clear illustration of this empowerment is the regional decision to implement DOTS, despite the lack of enthusiasm and even outright opposition on the central level.

Also, a few NGOs were included in the networks. However, when talking about civil society in post-Soviet states, certain modifications must be made. 'NGOs' have in communist countries originated in a political tradition contrary to the idea of civil society. Civil society in the Soviet bloc did not have the opportunity to develop independently of the State, for obvious reasons. With regard to empowerment and mobilization of new actors, we find that NGOs of governmental origin – of which there are quite a few in the post-Soviet countries – are less interesting than the very few but still active *actual* NGOs.[19] Some of these organizations have experienced great difficulties in their activities. For instance, a group that works to promote better conditions for prostitutes in St Petersburg is being more or less methodically countered by law enforcement officers. In receiving funding from international health collaborations (which it would not receive from its own government), this group has made a difference for sex workers in the city. That said, this is a rare example. Empowerment of Eastern NGOs is, however, clearly stated as a programmatic goal, although not a prominent feature, in health collaborations.

Region building

One feature of the efforts to build regional cooperative bodies in Europe in the early 1990s was the rhetorical emphasis on more or less well-founded perceptions of a common Northern cultural heritage. This was the case in the establishment of the BEAR, embracing the idea of the pre-revolution Pomor trade between Northwestern Russia and the Nordic countries, and the Euro-Mediterranean Partnership (EMP), building on *Mare Nostrum* rhetoric.[20] Equally, the Hansa trade and *Mare Balticum* were predominant in the establishment of the CBSS partnership.[21] Even in the area of international health collaboration, in the case of the Task Force, we can see that it relied on the *Mare Balticum* rhetoric in its self-justification:

> If we could step back a thousand years and take a bird's eye view of the Baltic Sea and the Barents region [. . .] we would see an area without other boundaries than those arranged by Nature herself. The Baltic Sea linked people together, but divided nations. Yes, Vikings crossed the sea, but most seafarers were merchants. People of diverse backgrounds interacted and civilisations influenced each

other. Yes, there were brutal wars, but they were but dark inter-
ludes in the evolution of peaceful co-existence.

(Task Force 2002: 5)

There is an evident, slightly romantic attempt at downplaying what
divides and stressing what unites. While the CBSS and the Task Force
itself have only recently sprung into existence, they are seen as rest-
ing upon a strong historical foundation. It was the disparity between
Eastern poverty and Western wealth, Eastern epidemics and Western
health that paved the way for an ambitious attempt to make health
the object of international politics in the Baltic Sea Region. A 'soft
security' perspective was prominent in this initiative, as the Task Force
was established to prevent communicable diseases from becoming a
destabilizing factor in the area. Rhetorically, the initiative rested upon
a notion of common values, history and culture in what was called the
Mare Balticum area.

Have health collaborations contributed to the 'building' of the Arctic region? Our answer to this question is clearly 'yes'. Admittedly,
there is a methodological challenge in assessing the wider contribution
of the health programmes since it is difficult to say what the situation would have been without them. Also, the goal of contributing
to 'more' integration is not very ambitious, so it actually adds little
meaning to say that 'yes, the initiative has engendered wider contacts'.
Without embarking on a hypothetical discussion of the impact of the
health programmes as compared to what would have happened without them, we can say that yes, they have enhanced East–West contact
in the region, they have enhanced contact between Russia and the Baltic states in the eastern part of the region and even enhanced contact
within individual countries (between the federal subjects of the Russian
Federation). The various levels of the Task Force structures have had
formal meetings three to five times a year during the mandate period
(Rowe 2004: 4–6). In addition, there has been substantial communication concerning projects both in the Task Force and the Barents Health
Programme.

We see this first, and most conspicuously, in the formation of a large
number of networks involving politicians, civil servants and, in particular, medical personnel linking Eastern and Western countries around
the Baltic Sea. Second, although probably less intended by the initiators
of the programme, the health programmes have brought Russia and the
Baltic medical communities closer together. After a decade of animosity following the break-up of the Soviet Union, the Arctic regimes have
contributed to re-connecting old networks, primarily in the medical

sector but also more widely. Relations were especially strained after the (short-lived) Soviet military intervention in the Baltic capitals in 1991, but have since developed very positively. As a St Petersburg interviewee said: 'an outsider [the Task Force] was needed for this to happen.' Third, and also probably unexpectedly from the initiators' point of view, the Western-generated programmes established contacts between the various federal subjects *inside* Russia. Contrary to what many in the West may assume, ties between neighbouring regions like Murmansk and Arkhangelsk oblasts in Northwestern Russia have traditionally been quite weak. Each has related primarily to federal authorities in Moscow and, though to a lesser extent, to the 'regional capital' of St Petersburg. Trans-national initiatives such as the Task Force, and more widely the BEAR, have provided arenas where representatives of the different Russian regions can make contact and share experiences. In sum, region-building efforts under the health initiatives were initially quite successful, and – most interestingly – what has taken place *inside* the eastern part of the region is actually as important as what has been achieved at the level of East–West relations.

Networks and sustainability

The mobilizing and region-building effect of the networks was seen by the participating states as something worth developing and broadening. The Northern Dimension Partnership was meant to do just that. The structure and aims of the Partnership were, as we have noted, based largely on the Task Force model, and three of the Task Force programme groups (HIV/AIDS, primary health care and prison health) have been incorporated in its structure as 'expert groups'. In addition, a new 'expert group' on non-communicable diseases has been established. The mere incorporation or establishment of expert groups, however, is not proof of continued activity. To keep the networks alive, there is an obvious need not only for structure, but also for motivating content – new tasks for the networks to solve, in order to justify the time and costs of active participation. Although representatives of the three expert groups formally incorporated into the Partnership maintained that much remains to be done and expressed the desire to continue their activities, it is at present (February 2006) highly unclear if they will be able to do so. So far, the 'expert groups' have not been given any tasks, and funding for their meetings has been scarce. One exception is the HIV/AIDS group, which has acquired money through an alternative channel, the Finnish agency STAKES (the National Research and Development Centre for Welfare and Health) as part of

Finnish work under the Barents Health Programme (Rowe and Rechel 2006).

Thus, it is fair to say that even though well-functioning networks were created under the auspices of the Task Force, they are not hardy or sustainable enough to be what had been hoped for: a self-sustaining undertaking after the programme was terminated. However, there can be no doubt that the Task Force has demonstrated the need for and rewards of mobilizing formerly isolated groups such as regional Russian medical personnel in cross-border relations. This is of particular importance in matters concerning communicable diseases. Therefore, it seems remarkable that the programme specifically designed to continue and develop this achievement – the Partnership – has not been able to fulfil this task (Rowe and Rechel 2006).

Conclusions

The causal role of Arctic regimes in bettering public health in the region is difficult to assess. Of course it could be argued that any intervention in a structure considered to be on the verge of complete breakdown would be a contribution towards something better – but quantifying such a change is a far too complicated task for this chapter to address.

What we *can* say is that the health interventions described in this chapter have resulted in some behavioural adaptations. They have contributed towards a new and more effective control of communicable diseases. Russian Arctic public health services have been working to implement the DOTS strategy, despite the opposition voiced by members of the centrally placed medical establishment and by officials in the Ministry of Health, although a change has taken place also here (the latter must be credited primarily to the WHO Moscow office). There is little doubt that the specific programmes of Arctic regimes that have promoted DOTS have spurred this development. In purely behavioural terms, the changes in Russian Arctic regions like Murmansk and Arkhangelsk oblasts are substantial: instead of costly mass screening of the population, self-reporting patients are being diagnosed through microscopy testing of sputum smear. Instead of long stretches of hospitalization and isolation, patients are now recuperating during treatment in their own homes. Instead of inaccurate long-term medication, patients are given short-term standard cures. All medicine is handed to the patient at the dispensary and is consumed under observation. If the patient fails to present himself or herself for medication, he/she will be located.[22] The DOTS strategy is definitely a more cost-efficient approach than what was the case with the Russian traditional diagnostics and treatment. Avoidance of long hospitalization spells, mass screening and inefficient medication has made this tuberculosis strategy work in a crisis-ridden Russian health care system. On a more general note, the health programmes have contributed to a wider debate within post-Soviet countries, first and foremost Russia, about strategies for countering the spread of communicable diseases.

Can we say that the Arctic regimes through their medical programmes have duplicated other activities? Or have they addressed issues that needed attention? First, there can be little doubt that much needs to be done in the field of public health. Thus, the need on the ground for the activities of the Barents Health Programme, the Task Force and the Partnership is not disputed. Furthermore, all programmes have built on existing bilateral projects, and are therefore not competing with these. Instead, they have concentrated on broadening the scope of these projects as well as coordinating them by providing a common framework. Thus we may conclude that any effect these programmes may have had on coordinating health efforts in the countries and regions concerned has been positive. To a large extent, the programmes also have concentrated their efforts in the Russian regions, as opposed to the centre, which can be seen as a division of labour between them and the WHO.

In a sense, the functional broadness of the Arctic regimes has been of significance to the success of the health programmes. Cross-border relations have been established, and experiences from other areas of activity have contributed to the various countries' accumulated knowledge of one another and how to run concerted efforts. The mere fact that both the CBSS and the BEAR had existed for several years before the inception of the health-related programmes provided a useful framework. Additionally, this has facilitated – at least rhetorically – the necessary political commitment from all participating countries. On the other hand, we have seen that rhetorical support to a programme has not always been followed through. Especially in the case of the Task Force, despite enthusiastic and supportive statements from all governments, only Norway contributed substantial funding (Rowe 2004). The lack of financial support from major powers like Germany does not, however, seem to have halted the general ambition to implement health projects as foreseen. Therefore, we can say that in these geographically limited health programmes (basically the Baltic states and Northwestern Russian regions), the enthusiasm and funding of smaller powers has been sufficient to keep the work going. On the other hand, this is not necessarily the case with the latest addition to the Arctic health regimes, the Partnership, which aims – much more than the two others – at unifying the wishes of all participant states in one coordinating body. Its focus is more directed towards the central political milieu than was the case in the Task Force and in the Barents Health Programme. The future of the Partnership seems far more dependent on its reception in all capitals, and most importantly in Moscow and in Berlin.

We must say that the programmes have contributed towards a more integrated region. Despite the weaknesses of the networks demonstrated above, some degree of mobilization and, connected with this, region building has indeed taken place. Most notably, regional medical workers and bureaucrats in Russian oblasts have, by participating in projects, been able to connect not only with their Western counterparts but, perhaps more importantly, with their colleagues in other oblasts. Similarly, integration between Russia and the Baltic states has been promoted as a direct result of the networking that has taken place, especially under the auspices of the Task Force.

That wider contacts can be said to have had any effect on the discursive regionality in the health sector, in the sense of a firmer feeling of oneness is, however, less likely. The gaps between East and West in the field of public health remain enormous, with fundamental differences in both the qualitative level of services provided, and the underlying philosophy of how, why and to whom to render these services. We see this in the promotion of DOTS, which is a strategy developed for third world countries, in Russia and the Baltic states. This illustrates the gap between standards in the public health services of the countries involved, and has, as noted, not been without friction. Although participants may have become more familiar with 'the other', this familiarity may well lead to a greater appreciation of the critical differences between the countries in the region, rather than a feeling of oneness.

Notes

1   A complete list of all interviewees can be found in Hønneland and Rowe (2004) and on file with the authors. Most interviews were carried out at the workplace of the interviewee and lasted from an hour to an hour and a half. All interviews were conducted without interpreter; most importantly, all interviews with Russians were conducted in Russian.

2   The following section describing some aspects of the health situation in post-Soviet states in the late 1990s is based on Hønneland and Rowe (2004: 1–3).

3   All statistics on morbidity and mortality in the Baltic Sea region are cited from Task Force (2000: 9–11). A scientific study of the health consequences of the collapse of the Soviet Union is found in McKee (2001).

4   The following presentation of the Barents Health programme is based on Hønneland and Moe (2002).

5 For a detailed presentation of the Task Force's inception, initiation and organization, see Hønneland and Rowe (2004: 21–46).

6 The Task Force consisted of the eleven member states of the Council of Baltic Sea States (Denmark, Estonia, Finland, Germany, Iceland, Latvia, Lithuania, Norway, Poland, Russia, and Sweden) and the European Commission.

7 The following presentation of the Northern Dimension Partnership in Public Health and Social Wellbeing is based on Rowe and Rechel (2006).

8 The eight member organizations in the Partnership are the Barents Euro-Arctic Region, the Council of the Baltic Sea States, the European Union, the International Labour Organization, the International Organization for Migration, the Organization 'Norden', UNAIDS and the WHO. See homepage of the Partnership Secretariat; http://northerndimension.custompublish.com/.

9 The first Northern Dimension action plan was adopted in 2000.

10 http://www.europa.eu.int/comm/external_relations/north_dim/ndap/ap2.pdf, accessed 24 December 2005.

11 Ibid.

12 http://www.ndphs.org/index.php?cat=29608, accessed 25 February 2006.

13 For more detailed information on the Partnership, see the homepage (note 8).

14 The following discussion is based on Hønneland and Rowe (2004, 2005).

15 This is reflected in our interview with an employee of the Moscow WHO office: 'We cannot compromise on DOTS. For us, this is a political matter.' However, within the WHO there have been some shifts on DOTS in recent years, and the organization has acknowledged that there are limitations to the strategy (see http://www.stoptb.org/stop_tb_initiative/ for more information on this). This trend, however, does not modify the Russian perception of the WHO as uncompromising in matters concerning DOTS.

16 The first steps were taken in 1994, when MERLIN (the Medical Emergency Relief International) established a collaborative project with the regional authorities of Tomsk Oblast in Siberia (Mawer *et al.* 2001). By 2003, DOTS had been introduced in 26 of Russia's 89 federal subjects (World Health Organization 2003: 105).

17 DOTS was introduced to, and largely implemented in, the Baltic countries under the No TB-Baltic programme, which was funded by the Nordic countries and subsequently incorporated in the Task Force structure (see Hønneland and Rowe 2004: 76–8).

18 Interview with Arkhangelsk-based Task Force official.

19 For a discussion of historical perspectives on civil society in Russia, see Volkov (2003). For a short discussion of the different types of 'NGOs' engaged in the Task Force, see Hønneland and Rowe (2004: 85–6).

20 For a discussion of the 'Barents rhetoric', see Hønneland (1998). The implications of the more extreme variant of this rhetoric – the 'Barents euphoria' – for East–West environmental cooperation in the European North are discussed in Hønneland (2003).

21   The Baltic Sea rhetoric is briefly discussed by Joenniemi and Stålvatn (1995: 9–11).
22   This is where several NGOs have taken an active part in countermeasures against tuberculosis. For example, the Russian Red Cross in Arkhangelsk is instrumental in locating missing patients and supplying them with medicine.

References

Barents Euro-Arctic Council (1999) *Health Co-operation Programme in the Barents Euro-Arctic Region 1999–2002*, Sixth Barents Euro-Arctic Council meeting, Bodø, 4–5 March 1999.

Coker, R. (2001) 'Control of tuberculosis in Russia', *Lancet*, 358: 435.

Field, Mark G. and Twigg, Judyth L. (eds.) (2000) *Russia's Torn Safety Nets*, New York: St. Martin's.

Hønneland, Geir (1998) 'Identity formation in the Euro-Arctic Barents Region', *Cooperation and Conflict*, 33: 277–97.

Hønneland, Geir (2003) *Russia and the West: Environmental Co-operation and Conflict*, London: Routledge.

Hønneland, Geir (2005) *Barentsbrytninger: Norsk nordområdepolitikk etter den kalde krigen*, Kristiansand: Høyskoleforlaget.

Hønneland, Geir and Moe, Arild (2002) *Evaluation of the Barents Health Programme – Project selection and implementation*, Lysaker: FNI report 7/2002.

Hønneland, Geir and Rowe, Lars (2004) *Health as International Politics: Combating Communicable Diseases in the Baltic Sea Region*, Aldershot: Ashgate.

Hønneland, Geir and Rowe, Lars (2005) 'Western versus post-Soviet medicine: fighting tuberculosis and HIV in North-West Russia and the Baltic States', *Journal of Communist Studies and Transition Politics*, 21 (3): 395–414.

Joenniemi, P. and Stålvatn, C.E. (1995) 'Baltic Sea politics: achievements and challenges', in P. Joenniemi and C.E. Stålvatn (eds.), *Baltic Sea Politics: Achievements and Challenges*, Stockholm: Nordic Council.

Kimerling, M.E. (2000) 'The Russian equation: an evolving paradigm in tuberculosis control', *International Journal of Tuberculosis and Lung Disease*, 4: S160–S167.

Mawer, C., Ignatenko, N.V., Wares, D.F., Strelis, A.K., Golubchikova, V.T., Yanova, G.V., Lyagoshina, T.V., Sharaburova, O.E. and Banatvala, N. (2001) 'Comparison of the effectiveness of WHO short-course chemotherapy and standard Russian antituberculosis regimens in Tomsk, Western Siberia', *Lancet*, 358: 445–9.

McKee, Martin (2001) 'The health consequences of the collapse of the Soviet Union', in D. Leon and G. Walt (eds.), *Poverty, Inequality and Health: An International Perspective*, Oxford: Oxford University Press.

Perelman, M.I. (2000) 'Tuberculosis in Russia', *International Journal of Tuberculosis and Lung Disease*, 4: 1097–1103.

Rechel, Bernd and McKee, Martin (2004) *Learning Lessons from the Experience of the Task Force on Communicable Disease Control in the Baltic Sea Region. Programme evaluation*, London: London School of Hygiene and Tropical Medicine.

Rechel, B., Shapo, L., McKee, M. (2004) *Millennium Development Goals for Health in Europe and Central Asia: Relevance and Policy Implications*, Washington DC: The World Bank (Working Paper No. 33).

Rowe, Lars (2004) *Report from the Steering Committee for Evaluation of the Task Force on Communicable Disease Control in the Baltic Sea Region*, Lysaker: FNI report 9/2004.

Rowe, Lars and Rechel, Bernd (2006) 'Fighting Tuberculosis and HIV/AIDS in Northeast Europe: sustainable collaboration or political rhetoric?', *European Journal of Public Health*, forthcoming.

Suhrcke, M., Rechel, B., Michaud, C. (2005) 'Development assistance for health in Central and Eastern European Region', *Bulletin of the World Health Organization*, 83: 920–7.

Task Force (2000) *Healthy Neighbours*, Task Force on Communicable Disease Control in the Baltic Sea region, Oslo: Ministry of Health.

Task Force (2002) *Report to the Heads of Government*, presented at the Baltic Sea States Summit by the Task Force on Communicable Disease Control in the Baltic Sea Region, St Petersburg, 10 June.

Task Force (2004) *Final Report to the 5th Baltic Sea States Summit, Laulasmaa, 21 June 2004*, Task Force on Communicable Disease Control in the Baltic Sea Region, Oslo: Ministry of Health.

Twigg, Judyth L. (2000) 'Unfulfilled hopes: the struggle to reform Russian health care and its financing', in Mark G. Field and Judyth L. Twigg (eds.), *Russia's Torn Safety Nets*, New York: St. Martin's.

Volkov, V. (2003) '"Obshchestvennost": Russia's lost concept of civil society', pp. 63–72 in N. Götz and J. Hackmann (eds.), *Civil Society in the Baltic Sea Region*, Aldershot: Ashgate.

World Health Organization (2003) *Global Tuberculosis Control: Surveillance, Planning, Financing*, WHO/CDS/TB/2003.316, Geneva: World Health Organization.

5 Pollution and conservation

Olav Schram Stokke, Geir Hønneland and Peter Johan Schei

Introduction

Environmental problems loomed large when the post-Cold War international institutions on the Arctic were created.[1] The ministers who established the Arctic Council decided to focus on matters common to Arctic states, 'in particular issues of sustainable development and environmental protection'.[2] They also resolved to 'coordinate and oversee' the programmes under the Arctic Environmental Protection Strategy, which had been established in 1991.[3] In the Copenhagen Declaration on the Council of Baltic Sea States, the foreign ministers noted 'that the establishment of closer cooperation between their countries creates better possibilities for solving jointly the environmental problems' in the region.[4] Similarly, the Kirkenes Declaration on the Barents Euro-Arctic Region made it clear that '[t]he objective of the work of the Council will be to promote sustainable development in the Region, bearing in mind the principles and recommendations set out in the Rio Declaration and Agenda 21'.[5]

These formulations reflect a belief that there are considerable gains to be reaped from coordinating measures for environmental monitoring and protection in the North. They were also spurred, however, by the fact that substantial collaboration was already under way in this issue area, and that might head-start broader initiatives (Stokke 1990). Governmental and other players in states bordering on the Soviet Union had seen improvements in East–West relations in the late 1980s as a window of opportunity for achieving progress in politically salient issues of trans-boundary pollution, especially from Northwestern Russia. Moreover, the process of creating new Arctic institutions coincided with the 'second environmental wave' in global politics, kicked off by the 1987 Report of the World Commission on Environment and Development and culminating with the 1992 UN Conference on

Environment and Development and the adoption of global conventions on climate change and biodiversity.[6]

This chapter sketches the environmental issues that have been pinpointed in Arctic fora and outlines Arctic institutional responses, including programmatic and normative activities.[7] Although some activities under the Council of Baltic Sea States are relevant to North-western Russia, notably those involving nuclear and radiation safety, this collaboration has otherwise been overwhelmingly oriented at areas outside the Arctic and will not be examined here. Applying the analytical framework advanced in Chapter 2, we then seek to trace the impacts of those activities on governments' abilities to solve or mitigate specific environmental problems – as well as on patterns of participation in Arctic affairs and regional connectedness. Two salient environmental issues are discussed in greater depth elsewhere in this book – the Arctic dimension of climate change in Chapter 6 and regional oil and gas activities in Chapter 7 – and will therefore not be examined in detail here.

The Arctic environmental agenda

Environmental conditions in the Arctic are highly varied, but some features are shared (AMAP 1997). The climate is marked by extreme variation in light and temperature, short summer seasons, snow and ice cover in winter, and large areas of permafrost. The coastal zone, where most of the population, industries and military operations are to be found, is the most sensitive to human pressures on natural resources and the environment. However, vulnerability varies immensely across the region. Some Arctic marine areas, including the Barents Sea, are among the most productive in the world, but low temperatures and little sunlight slow down the evaporation of toxic components and the physical, chemical and biological breakdown of pollutants. This may reduce the capacity of ecosystems to regenerate if significantly disturbed. Some terrestrial and marine ecosystems are rather simple, which can mean that disruption of one link in the food chain – for instance, through over-exploitation – will severely affect the rest of the system.

Aware of these general features but recognizing the fragmented and crude nature of the information available on various substances that might threaten the Arctic environment, policy makers placed environmental *monitoring* high on their list of priorities. The 1991 Rovaniemi Declaration that launched the Arctic Environmental Protection Strategy (AEPS) committed the eight Arctic governments to 'monitor the levels of, and assess the effects of, anthropogenic pollutants in all components of the Arctic environment'. Similarly, the Kirkenes Declaration lists 'expanded monitoring of ecology and radioactivity in the Region' as one of three priorities in the environmental sphere.

Yet another feature common to Arctic territories is that they serve as reservoir, or sink, for many *hazardous substances* generated and discharged elsewhere. Most of the radionucleides currently found in Arctic marine and terrestrial environments originate from activities undertaken outside the region (AMAP 2002: 59). Main sources are British and French reprocessing plants, atmospheric nuclear tests conducted four to five decades ago, and fallout from the Chernobyl accident in 1986. Also many other pollutants have been produced and discharged well outside the Arctic. Strong south–north air flows, rivers, and ocean currents transport a range of hazardous compounds into and within the Arctic. Particular worries concern persistent organic pollutants (POPs), including organochlorine pesticides used in agriculture, industrial chemicals such as polychlorinated biphenyls (PCBs), and various combustion products. The low temperatures in the Arctic serve as a 'cold trap' for some of these POPs and prevent further transport. Similarly, some heavy metals found in high concentrations in the Arctic, among them mercury, originate largely from waste incineration and coal-burning power plants and residential heaters as far away as Asia – and these are discharges expected to accelerate in the years ahead due to economic growth, especially in China (AMAP 2002).

The effects of POPs and heavy metals on humans are more dramatic in the Arctic than those documented at lower latitudes because such substances bio-accumulate in the fatty tissue and blood of some species, including marine mammals and sea birds, important in the diet of Arctic indigenous residents. The Inuit of Canada and Greenland have among the highest exposures to PCBs and mercury measured on the planet. Foetuses and small children relying on breast milk are particularly vulnerable (Dewailly and Furgal 2003). These hazardous substances also demonstrate the close links that sometimes exist between pollution and conservation issues. Some of the highest PCB levels ever measured in fat and blood are currently found in polar bears around Svalbard and Franz Josef Land, and recent studies suggest negative impacts on immune systems and reproduction (Reiersen *et al.* 2003: 76). Such effects are all the more dramatic because this is a species already threatened by declining ice extension, due to global warming, and by increased hunting activities.[8]

Important as external flows are, activities *within* Arctic states also generate a substantial share of regional pollution, including organochlorines, heavy metals, and hydrocarbons. Some of the largest and most heavily industrialized centres in Russia are found on the banks of rivers branching into the Arctic oceans. This is true for the Norilsk mining and metallurgical complex, the West Siberian oil and gas industries, the huge Kuznetsk coal basin, and even the nuclear reprocessing plant in Mayak, near Chelyabinsk on the southeastern slopes of the Urals. The Yenisei and Ob are the main channels for river-borne pollution into the Arctic. Similarly, as much as two-thirds of the atmospheric heavy metals found in the High Arctic originate from industrial activities in Northwestern Russia, as does most of the sulphur found within the Polar Circle (AMAP 1997: vii, 97–99). The smelters in the Kola Peninsula near Russia's Nordic borders and in Norilsk further east are major sources.

Attention to the external as well as the regional dimensions of Arctic environmental challenges was evident already in the founding documents of Arctic institutions and has been specified later. In the 1991 Rovaniemi Declaration, Arctic states resolved to 'take preventive and other measures directly or through competent international organizations regarding marine pollution irrespective of origin'. Six years later, they agreed to 'increase. . . efforts to limit and reduce emissions of contaminants into the environment and to promote international cooperation and make a determined effort to secure support for international actions in order to address the serious pollution risks reported by AMAP'.[9]

Unlike long-range transported pollutants, Arctic nuclear, hydrocarbon or shipping activities also involve leaks and risk of accidents that could imply very severe threats to the Arctic environment. Since oil, gas and shipping activities are addressed by Offerdal in Chapter 7, the emphasis here is on potential pollution from past and ongoing *nuclear activities*, especially in the European Arctic. Nowhere else on Earth is there such a concentration of civilian and naval nuclear reactors. In the years after the Second World War, the Soviet military complex appropriated vast land areas on the Kola Peninsula and built seven bases for its Northern Fleet. Neither the safety practices nor the quality of storages found in those bases for various types of nuclear waste has been reassuring. Until the late 1980s, high-level waste such as nuclear reactors still containing spent nuclear fuel was on several occasions dumped in the Barents and Kara Seas, and low-level liquid waste from cooling and incineration facilities of radioactive installations has been dumped even later (Stokke 1998; 2000). Russia's radioactive waste problem will accelerate in the years ahead since numerous nuclear submarines have now been taken out of operation in accordance with the

Strategic Arms Reduction Treaty regime. Radioactivity was among the five categories of pollutants prioritized in the Rovaniemi Declaration on the AEPS, and 'enhanced work on the operational safety of nuclear facilities' was one of three priority issues in the Kirkenes Declaration on the Barents Euro-Arctic Region (BEAR). In 1997, the eight Arctic ministers agreed to 'fully support regional cooperation between two or more Arctic States, as well as multilateral efforts, to enhance nuclear reactor safety and to increase and promote the safe management, storage and disposal of spent nuclear fuel and radioactive waste'.[10]

At a general level, most of the problems related to Arctic *conservation issues* were known and formulated long before the creation of Arctic cooperative institutions. Overexploitation of biological resources has been more the norm than the exception in the area: examples include polar bear, walrus, the bowhead and the Northern right whale, and on Svalbard the local sub-species of reindeer. Beginning in the 1970s, somewhat more conservation-oriented ideas gained influence in the regional management of living resources, as evident in the 1973 Polar Bear Agreement and the establishment of national parks and nature reserves in Svalbard and Northeast Greenland. In the USA, the Arctic National Wildlife Refuge in Alaska had been established more that a decade earlier, but the total area of the refuge was doubled in 1980 when the Alaska National Interest Lands Conservation Act was adopted.

The growing attention to conservation in the 1970s had been spurred by the dramatic collapse of the Atlanto-Scandian herring, and by greater political attention worldwide to environmental problems, most clearly articulated at the 1972 United Nations Conference on the Human Environment in Stockholm. Concurrently with the formation of new Arctic institutions two decades later, the Rio Conference on Environment and Development and the negotiation of the Convention on Biological Diversity (CBD) had introduced new normative standards for conservation, notably the linkage between conservation, sustainable use and benefit sharing in management of living resources and their habitats. On this background, it is not surprising that conservation and biodiversity issues were integrated right from the outset in Arctic institutions. Thus, the Rovaniemi Declaration singled out the conservation of Arctic flora and fauna as an important area of Arctic collaboration, although the initial emphasis was on 'exchange of information and coordination of research on species and habitats'. The Kirkenes Declaration went somewhat further, noting that 'the environmental dimension must be fully integrated into all activities in the Region, *inter alia*, through the establishment by the states in the Region of common ecological criteria for the exploitation of natural resources and the prevention of pollution at source'. The Barents Euro-Arctic Council (BEAC) Environmental Action Programme formulated an ambitious goal for 'green' conservation: 'Maintenance of the biodiversity of the Region and the natural quality of the pristine areas'.[11] Such maintenance is a daunting task, practically impossible to achieve throughout the Barents Region. Further, there is in this first programme little more specific on Arctic problems; and the references to international commitments are rather general, including the statement that states are to support 'implementation of the CBD in the Region'. Neither early Arctic Council documents nor those prepared under BEAR describe or discuss very well the specific problems related to biodiversity conservation in the Arctic. Such problems are identified and more or less accepted as being there and causing biodiversity to decline or deteriorate to various extents also in the Arctic.

A striking feature of the first years of BEAR collaboration was that questions about 'the environment' were supposed to permeate the entire collaborative venture, spanning from issues like culture and tourism to indigenous peoples and science (Hønneland 2003: 121–2). In later formal declarations from the major BEAR institutions, environmental concerns are clearly still on the agenda, but assigned somewhat less prominence than during the formative years. The introduction to the BEAC 10 Year Anniversary Declaration from 2003 mentions 'people-to-people cooperation' as a major achievement, whereas environmental concerns are listed alongside several other sectoral priorities further down in the document. There is now no reference to the environment as an overarching concern in the regional collaboration: it is one area of cooperation among several others. In financial terms, as noted in Chapter 4, health took over as the first priority when the Barents Health Programme was established in 1999.

Also the priority between different environmental questions has changed – from an initial focus primarily on nuclear safety and pollution from the Kola smelters, to biodiversity and water quality. This might reflect the fact that despite serious environmental degradation around the Kola smelters, the level of documented current impact on human health is low. Nor do POPs represent any serious danger to health and industry in this part of the Arctic; and radioactive contamination is low at present. Although in areas close to the smelters air pollution has been shown to cause health problems, the main environmental disease-generating problem in the Russian part of BEAR is the poor quality of the drinking water. Pollution of drinking water comes largely from the disposal of refuse close to the sources, and the lack of purification facilities. There is reason to believe that drinking-water quality was given increased attention once the Western members of the BEAR collaboration gradually became aware that this was a top priority on the Russian side.

To summarize, both pollution and conservation issues were high up on the agenda when the Arctic institutions were formed. In addition to climate change and regional oil and gas activities, which are dealt with in separate chapters in this volume, the environmental issues prioritized under Arctic institutions cluster around four agenda items that will structure our discussion of effectiveness: environmental monitoring, hazardous substances, nuclear safety, and conservation and biodiversity. The transition from the AEPS to the Arctic Council signalled an aspiration to integrate those issues with activities aimed to promote economic growth, under the broad label of sustainable development (Bloom 1999). In practice, as shown below, those aspirations have been most visible with respect to assessment work on environmental impacts of Arctic operations that are believed to increase in the years to come, including those related to oil, gas and shipping. To some extent, under BEAR the role of traditional pollution and conservation issues has changed in the opposite direction. In the early years, it was portrayed as a common denominator that would permeate all collaboration, whereas today it is one of several sectors and hardly the most conspicuous among them.

Arctic institutions at work

The Arctic Council

The Arctic Environmental Protection Strategy emerged in 1991 from a Finnish initiative as an intergovernmental vehicle for environmental cooperation.[12] The main mechanisms for putting this strategy into action were a set of permanent working groups responsible for four key areas: environmental monitoring, marine pollution, emergency preparation, and conservation. When formally subsumed under the Arctic Council in 1998, the AEPS groups were complemented by the Working Group on Sustainable Development.[13] In the two-year periods between ministerial meetings, programme activities are overseen by the Senior Arctic Officials (SAO), who represent the foreign ministries of the 'Arctic eight'.

Among the five working groups, the Arctic Monitoring and Assessment Programme (AMAP) has been coined the 'jewel in the crown' (Archer and Scrivener 2000). This programme organization aims to develop better knowledge about the sources, pathways and concentrations of regional pollution, as well as their impacts on human health and Arctic flora and fauna. Pollutants in focus are POPs, heavy metals, Arctic haze and acidification, radionucleides and hydrocarbons. These monitoring activities have gradually been expanded to include regional consequences of stratospheric ozone reduction and global climate change. Under this programme, two comprehensive AMAP Assessment Reports have been produced, the latest in 2002, but such high frequency in issuing thematically comprehensive assessments will not be maintained in the future.[14] As evident in the most recent assessments, on health (2002), climate change (2005) and oil and gas activities (2006), the trend is towards thematically more specific products. Reports are sometimes produced in collaboration with other Arctic Council working groups, especially that on Conservation of Arctic Flora and Fauna (CAFF).[15] Comprehensive assessments are now envisaged every 10 years, with specific and thematically flexible Issues of Concern reports to be produced for each biannual ministerial, or if appropriate for the SAO meetings held in-between (AMAP 2004: 4). Under CAFF, a Circumpolar Biodiversity Monitoring Programme was recently launched, closely linked to the UNEP World Conservation Monitoring Centre; it promises to become a useful tool for following the status and trends of Arctic biodiversity.

Besides monitoring and assessment, Arctic Council working groups engage in evaluation of policy priorities, soft norm building, and capacity enhancement activities. With respect to evaluation, the Working Group for Protection of the Arctic Marine Environment (PAME) initially focused on the adequacy of existing international legal instruments for marine protection. The backdrop was certain disagreement among the Arctic states on the appropriateness of new international regulatory measures that specifically addressed the Arctic region (Stokke 1992). Subsequent evaluations undertaken by this working group have focused on environmental hazards related to increased regional shipping activities, the most ambitious of which is the Arctic Marine Shipping Assessment, to be completed in 2008.[16]

Yet another working group, tasked with Emergency Prevention, Preparedness and Response (EPPR), early on produced an environmental risk analysis of Arctic activities.[17] Until recently, the main focus of this working group was on petroleum activities, with the Circumpolar Map of Resources at Risk from Oil Spills in the Arctic (EPPR 2002) as an important output. Today the scope has been expanded to include nuclear installations and other facilities that use or store hazardous material, and natural disasters.

The normative activities of Arctic Council working groups are modest in that few efforts are made to create rules that are more ambitious or more specific than those already embraced in broader international fora. A recurrent activity under PAME, for instance, has been to encourage Arctic states to sign and ratify international conventions of particular relevance to regional marine pollution.[18] More substantively, this working group also engages in the development of Arctic guidelines for certain activities that pose threats to the Arctic environment, sometimes jointly with the EPPR. The most prominent example is the 1997 Arctic Offshore Oil and Gas Guidelines, reviewed and updated in 2002.[19] More specific products have been the Field Guide for Oil Spills Response in Arctic Waters and the Guidelines for Transfers of Refined Oil and Oil Products in Arctic Waters.[20] Under CAFF, the International Murre Conservation Strategy and Action Plan and the Circumpolar Eider Conservation Strategy and Action Plan provide detailed suggestions for protecting those species and the breeding and feeding areas they depend upon.

Finally, Arctic Council working groups engage in capacity enhancement, with a strong focus on Russia. Under the Arctic Council Action Plan to Eliminate Pollution in the Arctic (ACAP), various internationally funded projects have sought to improve the collection, storage and decontamination of various hazardous substances – especially PCB and obsolete pesticides. Under PAME's Regional Programme of Action to Eliminate Pollution from Land-Based Activities, several projects have aimed at developing and implementing a Russian Plan of Action.[21] A more diffuse kind of capacity enhancement, oriented toward information exchange and mutual learning, has been in focus in CAFF's Circumpolar Protected Area Network (CPAN), once considered a cornerstone of CAFF activities. A 1996 CPAN Strategy and Action Plan described existing and proposed protected areas in the Arctic and identified gaps and challenges. However, none of the eight countries has been willing to take the lead in this area, and the development of CPAN has been lagging behind.

Sub-regional cooperation: the BEAR

The Environmental Task Force of the BEAR cooperation was established in 1994 to advise the (minister-level) Barents Euro-Arctic Council on objectives, priorities and actions for environmental cooperation in the region. In 1999, its name was changed to the BEAC Working Group on the Environment. It meets one to three times a year between the meetings of the region's environmental ministers. Main priorities have been to identify environmental hot spots in the region, and to introduce cleaner production schemes in Northwest Russian industry, biodiversity and sustainable forestry, and clean drinking water in the Russian part of the region.[22]

Under the (province-level) BEAR Regional Council, a working group on the environment was set up in spring 1993. The main task of the regional working group was to draw up an action programme for environmental issues in the region. Along with most other working groups under the Regional Council, it was dissolved in 1999 for cost reasons, but was re-established in late 2001 with EU funding.[23] The two main priorities of the working group have since been water quality and biodiversity, and its work is organized under the 'Barents 2010' umbrella. This is an Interreg-financed project aimed at elaborating strategies for the BEAR cooperation within industry, higher education, the environment and communication, managed by the Swedish county of Västerbotten. The regional working group is also working actively to link nature conservation and tourism in the Barents region.

Although as noted BEAR conservation goals have been lofty and partly unrealistic, actual work has been down to earth, especially in recent years – as shown, for instance, by the International Contact Forum on Habitat Conservation in the Barents Region. Established in 1999, this forum has been tasked with developing concepts and strategies to promote research and protection of different habitats. The contact forum has given special emphasis to the protection of old-growth forests, wetlands and marine habitats. Support to networks of nature protection areas is prioritized, as is capacity building for Russian nature protection institutions. This involves some norm development, especially with respect to sustainable forestry, but basically the normative content is taken from broader instruments and arenas, especially the CBD and the UN Forum on Forests.

The BEAR collaboration never evolved into the multilateral project-administrative structure anticipated by its initiators. Concrete projects, in the environmental sphere as is other sectors, have been organized mainly at the bilateral level between Russia and the individual Nordic states. Hence, the BEAR environmental working groups at both central and regional level have not financed or administered project implementation in the region to any significant extent, but rather attempted to coordinate the activities of environmental protection authorities in the region as well as various initiatives financed by other international programmes, primarily under the EU, such as Tacis and Interreg. As expressed by a former chair of the regional working group on the environment:

The first obstacle we met was in finding financing for our work. There was no common financing of the activities, as was the case with the North Calotte cooperation. This was disappointing for us. Instead, we had to – and still have to – apply for money in different countries for each project. This made multilateral projects difficult to carry out. Therefore, most of the projects in the Barents Region have been bilateral. The Nordic Council of Ministers has supported some of our multilateral work and this has been a great help.[24]

Financing for bilateral projects has come partly from the regular budgets of national or regional authorities, partly from 'Barents institutions' such as the Nordic countries' Barents secretariats. As mentioned, various EU programmes have also been important sources of finance for environmental projects in the region, as have international finance institutions such as the Nordic Environmental Finance Corporation (NEFCO). Individual projects are often supported by different sources, and it is a question of interpretation whether or not they should be placed under the BEAR umbrella. On the one hand, they take place within the BEAR core area and are coordinated and partly shaped and financed by 'BEAR money', for instance from the national Barents secretariats. On the other hand, the BEAR working groups on the environment are not multilateral project-administrative institutions which initiate and finance concrete projects. Compared to other sources of financing, contributions from national Barents secretariats have been quite limited. In late 2005, Norway took the initiative to transform the national Barents Secretariat in Kirkenes into an international secretariat for the entire multilateral collaborative arrangement, which – if implemented – might change this situation. As in other areas of regional cooperation in the European Arctic, however, donor countries seem reluctant to leave their project money to a multilateral institution – especially one closely associated with another country – and prefer instead to run their projects bilaterally and report them as (in this case) BEAR projects afterwards.[25]

Coordination

The activities of the many environmental working groups, sub-groups and expert teams that have emerged under the Arctic Council and the BEAR are many, substantively diverse, and typically highly specialized. With such activities overseen primarily by generalist SAOs, whose daily work is in foreign ministries rather than research laboratories or environmental agencies, it is not surprising that over time, the working groups developed a considerable degree of autonomy in identifying thematic foci and shaping programme direction. A review commissioned by the SAOs criticized structures and procedures of the Arctic Council, noting that working groups lacked coherence and awareness as to their respective activities (Haavisto 2001). Reports produced by the working groups now indicate a much greater orientation towards activities that cross institutional boundaries, including more joint endeavours and regular representation at each others' meetings. The chairs of all Arctic Council working group now come together on a regular basis to keep updated on activities and, as appropriate, to identify joint projects.

Coordination between the circumpolar and the sub-regional levels of governance is less systematic, although here as well there is occasional attendance by observers from Arctic Council working groups at corresponding BEAR meetings, and vice versa.[26] For instance, there does not seem to be any clear division of labour between CAFF activities and those undertaken under BEAR – nor between the various Arctic programmes and those undertaken under the Convention on Biological Diversity.

Environmental effectiveness

What is the primary value that Arctic institutions have added to ongoing efforts to solve the specific environmental problems targeted? Applying the analytical framework advanced in Chapter 2, this section evaluates the impacts of the activities described above on the four environmental priority areas pinpointed by Arctic institutions: environmental monitoring, protection from hazardous substances, nuclear safety, and protection of Arctic biodiversity. We will also consider the interplay of Arctic institutions with broader cooperative processes in these areas.

Improving environmental monitoring

Extensive systems for monitoring both air pollution and radioactivity have been set up since the establishment of BEAR. This has occurred largely at the bilateral level between Norway and Russia, partly including Finland as well, and at AMAP level with respect to the eastern part of the Barents Region.[27] The monitoring itself is mainly taken care of and financed by national research institutes. A Finnish-Norwegian-Russian project to monitor the effects of the planned modernization of the Pechenganikel smelter has received financing from several institutions, including the Norwegian Barents Secretariat. As with many other projects, the major source of finance is the EU, through its Inter-reg programme.

Despite such sub-regional activities, it is the Arctic Council that has been central in international efforts to coordinate the generation and analysis of data relevant to circumpolar environmental assessment. It is a general challenge in effectiveness analysis to estimate what would have happened in the absence of any particular institution, but it does seem that the Arctic Council has influenced the level and direction of regional monitoring and research activities conducted by Arctic as well as non-Arctic states. Such influence has been achieved in part by means of normative commitment. Neither the AEPS nor the Arctic Council has involved legally binding commitments, but there is no doubt that the high priority accorded to environmental monitoring in the 1991 Rovaniemi Declaration implied strong expectations within and outside the Arctic that governments would intensify such activities. For the Arctic states as a group, environmental monitoring has stood out as an attractive object of cooperation – in part because it does not raise controversial questions about the appropriateness of international regulation, but also because the benefits of harmonizing data collection and analysis throughout the circumpolar area are substantial, especially given the high costs of conducting environmental research in the Arctic. In tandem, those factors have created incentives for national institutions responsible for financing or conducting environmental research to allocate more resources to issues of concern to the Arctic Council. This has been especially true for states (including Canada and Norway) which had either initiated these cooperative processes and were thus eager to see them succeed, or had pushed the monitoring issue with particular vigour.

Although other working groups too have contributed to this outcome, coordination under AMAP has been the primary instrument. The basis for the various authoritative reports produced by AMAP on the state and dynamics of the Arctic environment is data acquired through national and international research and monitoring programmes. AMAP's main role has been to harmonize ongoing activities, by coordination and review of National Implementation Plans in light of the AMAP Trends and Effects Programme, and to promote studies and monitoring activities to close identified knowledge gaps. In the case of POPs, for instance, Denmark was induced by AMAP to upgrade existing ad hoc investigations to a systematic and long-term monitoring programme (Reiersen *et al.* 2003: 64). The Persistent Toxic Contaminants project under AMAP, with funding from the Arctic states and several international institutions, engaged the Russian ministries for natural resources and health as well as the Russian Federal Service for Hydrometeorology and Environmental Monitoring (Rosgidromet) in assessment of the significance of aquatic food chains as pathways of exposure for indigenous peoples, the relative importance of local and distant sources, and the role of atmospheric and riverine transport.[28] Similarly, the USA, which had provided very little data on POPs for the first AMAP assessment, intensified its collection of Alaska data for the second assessment, partly in collaboration with Canadian colleagues (Huntington and Sparck 2003: 221). However, the density of monitoring stations and sampling sites varies considerably across the circumpolar range, and this uneven distribution remains a major challenge. In order to strengthen and streamline reporting and use of relevant information, five AMAP Thematic Data Centres have been set up – on atmospheric, marine, freshwater and terrestrial, radioactivity and health information.[29]

In summary, although much environmental monitoring would probably have occurred even without the cooperative Arctic institutions, the Arctic Council has generated political commitments and incentives to intensify activities in this area and helped to harmonize methods and coordinate studies. Sub-regional monitoring activities have not been implemented, or to any significant extent financed, by BEAR institutions, but presenting initiatives as 'Barents projects' might have made it easier to attract financing from international programmes and national governments. All considered, there is no doubt that Arctic cooperation has improved the quality of data input and its utilization.

Combating regional hazardous substances

The fact that long-range transported POPs and heavy metals occur in very high concentrations in the fat of Arctic animals, and in the breast milk of some Inuit women in Canada, had been discovered prior to the establishment of Arctic cooperative institutions in the early 1990s (Dewailly and Furgal 2003). It was also known that this might impair immune systems and reproduction, and cause neuro-behavioural disorders. Seeking international action, Canada had presented these findings and raised the POPs issue under the Convention on Long-Range Transported Air Pollution (CLRTAP) already in 1989 (Selin 2000: 90). Subsequent systematic investigations under AMAP have extended sampling to the entire circumpolar area and clarified the pathways and mechanisms that are involved (Reiersen *et al.* 2003). The Arctic Council in particular has not only facilitated such investigations, but also enhanced their prominence in the public debate within the region and beyond.

Such awareness raising about the Arctic dimension of certain pollution problems is one significant impact of the institutions examined in this book. This has been achieved also because of the premium placed on dissemination activities. The AMAP Assessment Reports and the ACIA Report have been issued also in non-technical 'overview' versions that target decision makers and the general public; they can – like most other reports and background material – be downloaded from the AMAP website.[30] A series of fact sheets have been prepared on the most hazardous substances, explaining in non-technical terms their origins, use and associated threats. Similarly, the Circumpolar Map of Resources at Risk from Oil Spills portrays in an easily comprehensible way the stakes that are involved in Arctic petroleum activities.

Although consensual knowledge and political awareness are important ingredients in problem solving, it is only when they trigger action that actually helps to reduce discharges of hazardous pollution, or risks thereof, that we may speak about effectiveness. Some such action has clearly been taken under Arctic institutions. The most tangible instances concern the use and storage of hazardous material found within the region. A set of ACAP projects aimed at eliminating PCBs in Russia has, in addition to proposing legislative measures and outlining strategies for introducing substitutes, identified existing stockpiles of this material and collected large amounts for safer storage and ultimate destruction. Envisaging application of state-of-the-art Western technology, this Arctic Council endeavour has now moved to demonstrations of how PCB can be destroyed. A three-year proposed plan developed in 2002 involved the destruction of nearly ten percent of the PCB believed to remain in Russia, but four years later implementation still remains limited by the inability to identify, and receive permission, for an appropriate site for destruction facilities.[31] The NEFCO provided funds for some of these PCB activities, whereas other parts have been financed directly by Arctic states, especially the USA and Norway.[32]

Pesticides is another area in which the Arctic Council has spurred practical problem-solving efforts. Having identified a number of priority regions where obsolete or prohibited pesticides might have considerable effect on the Arctic environment, several ACAP projects have enabled inventorying of most stocks and repackaging and safe storage of more than a thousand tonnes of pesticides, with destruction as the next step.[33] Training programmes on cleaner production, tailored for engineers at the metallurgical complex in Norilsk, are expected to yield substantial reductions in energy use and emissions of carbon dioxide and nitrogen oxides (ACAP 2003). Other ACAP activities target mercury, dioxins and furan, but these have not proceeded beyond fact finding on releases, concentrations and cleaner production options and the identification of pilot projects (ACAP 2005). The same is true for the most recent substance addressed under ACAP – brominated flame retardants, which originate in electric and electronic equipment, insulation material and transport vehicles. A project plan was adopted in 2004, after the Second AMAP Assessment had found that those compounds were occurring in rising concentrations in the Arctic environment.[34]

The various types of operational guidance provided under the Arctic Council have a clear potential to reduce the risk of regular or accidental discharges of hazardous substances into the Arctic environment. The foremost instances are the approved guidelines, including the Offshore Oil and Gas Guidelines and the more specific guides for oil-spills response and oil transfer. Less formal but serving a similar function is the Manual on a Shoreline Cleanup Assessment Technique (Owens and Sergy 2004), adapted for Arctic conditions and meant to enable residents and decision makers to draw an accurate picture of the nature and extent of shoreline oiling following a spill, which is one necessary condition for effective response.

Explicitly non-binding, these various documents are intended to support governments and operators of Arctic installations and vessels in efforts to enhance environmental safety, pinpointing and responding to conditions that are specific to this region – such as low temperatures, ice presence and long periods of darkness. They provide information about the practices of Arctic states considered to be the most advanced on the various issues in question, and to some extent articulate norms developed under broader international organizations, such as the International Maritime Organization or regional pollution regimes like the OSPAR Convention on marine pollution. The normative force of these various guidances is low, however, and there is no systematic review of whether governments or others actually make use of them. Offerdal argues in Chapter 7 that such use seems to be very limited in the case of the Offshore Oil and Gas Guidelines. Given the largely informational character of such guidance documents, it is not surprising that dissemination is a priority. For instance, a thousand copies of the Field Guide for Oil Spills Response in Arctic waters were distributed worldwide to governments, oil companies, libraries and oil-spill cleanup contractors.[35]

To summarize: awareness raising, operational guidance, and transfer of equipment and technology clearly have a potential to impact on regional releases of hazardous substances, but actual effects have so far been limited, for several reasons. First, hazardous substances in the Arctic had emerged as an international issue before the emergence of the AEPS, although the latter added to its prominence. Second, the normative guidance provided by Arctic institutions is 'soft' and without any systematic review of actual application. Third, even the most advanced capacity-enhancement projects have been experiencing severe difficulties of implementation. Finally, as we noted earlier, major sources of Arctic pollution are found further south. This brings into focus the interplay of Arctic institutions with broader arenas for environmental governance, to be examined below.

Enhancing nuclear safety

Radioactivity was given priority right from the outset in the AEPS and results from AMAP monitoring activities have been fed into various international fora where measures to enhance nuclear safety in Russia have been discussed, including the Contact Expert Group established under the International Atomic Energy Agency in 1996. There has also been some recent Arctic Council activity on preparation for emergency situations at nuclear installations, but the sub-regional level of governance has been far more important regarding practical measures to enhance nuclear safety in the region.

Nuclear safety was a main environmental priority in the Kirkenes Declaration: indeed, this is probably the environmental problem that has attracted the most interest from abroad to Northwestern Russia. As with monitoring, however, the BEAR institutions have not been the major players. Shortly after the establishment of the BEAR, nuclear safety on the Russian side of the border rose to an almost exceptional prominence on Norway's foreign-policy agenda. In 1994 a White Paper on the issue was presented to the Storting (the Norwegian Parliament), which in turn instructed the government to establish a Plan of Action for nuclear safety in 'areas adjacent to our borders'. That plan was launched in 1995; over the next decade around USD 160 million was spent on nuclear safety projects in the post-Soviet states, mainly in the northwestern corner of Russia. Projects have varied from research and monitoring to large-scale construction initiatives like transport, storage and treatment facilities for radioactive waste. While projects under the former category have generally been considered successful (Hønneland and Moe 2000), construction projects have to a much larger extent suffered from disagreement and conflicts of interest between Norwegian and Russian players, and among various institutions within Russia. Nuclear safety has most certainly improved since the early 1990s – but this is mainly a result of activities outside the formal

BEAR arrangement. Nuclear safety projects are implemented partly by regional authorities in Norway and Russia, and the initiatives might also in this case be perceived by both actors and observers as 'Barents projects', but the institutional and financial links are closer to the Norwegian Plan of Action and to the Arctic Military Environmental Cooperation (AMEC) than to the BEAR.[36] Mention should be made of one significant exception: the BEAR was instrumental in bringing about the Multilateral Nuclear Environmental Programme in the Russian Federation (MNEPR), signed in 2003.[37] This agreement provides an institutional framework for international cooperation concerning the safety of spent nuclear fuel and radioactive waste management in the Russian Federation. Notably, it ensures indemnity against liability for foreign partners.

Conservation and habitat issues

There is no doubt that monitoring, networking and issue-specific strategy development under the CAFF and the BEAR to some extent build capacity among those involved to address species and habitat challenges in their domestic settings. The BEAR International Contact Forum on Habitat Conservation is deemed useful by those who participate; and Arctic Council outputs like the strategies and action plans on murre and eider are of high quality, cover conservation issues not well covered elsewhere, and appear practical enough to be implemented. It is equally clear that neither the BEAR nor the Arctic Council have been arenas for substantial strengthening of international conservation commitments or for diffusing powerful approaches to biodiversity management or raising substantial funds for conservation purposes. One reason is that many conservation and habitat issues do not have clear trans-boundary components. Moreover, each individual Arctic country is engaged in a wide set of international activities, policies and processes on biodiversity management in the Arctic, of which the Arctic Council and BEAR are hardly the crucial ones. Species whose population numbers have begun to increase, like the walrus and the barnacle goose, and the Svalbard reindeer, have been subject to long-lasting conservation efforts, so it is still too early to tell whether the Arctic Council or the BEAR has played a crucial role in saving any species yet.

Institutional interplay

The links between Arctic institutions and broader international endeavours are salient in the issue areas examined above. The role of the Arctic as a sink for long-range transported marine and airborne pollution was one of the reasons why PAME (1996), in its report to the Inuvik ministerial meeting, concluded that binding Arctic agreements were not a priority, since many legal instruments were already in place for major issues of concern – but lacked the participation of some important Arctic states. Moreover, in those areas such as POPs where existing rules were either absent or inadequate, creating or strengthening normative commitments must involve also states beyond the region, if one is to cover the main sources. These considerations highlight the significance of catalytic activities, i.e. efforts under Arctic institutions aimed at influencing the normative contents of, or participation in, other and usually broader institutions.[38]

Thus, in the case of atmospheric pollution, the AMAP monitoring network includes measurement stations that serve the European Monitoring and Assessment Programme under the CLRTAP. Monitoring activities under the Arctic Council contribute to reviews of the effectiveness of the CLRTAP Århus Protocols on POPs and heavy metals and of the Stockholm POPs Convention (AMAP 2004: 4). The AMAP Thematic Data Centre for marine information is located at the International Council for the Exploration of the Sea, which coordinates assessment of fish stocks and the marine environment in the North Atlantic.[39] As noted, the CAFF Circumpolar Biodiversity Monitoring Programme is placed within the UNEP World Conservation Monitoring Centre. Such nesting within broader programmes is advantageous not only for cost-efficiency reasons: it also ensures that Arctic monitoring and research efforts benefit from expertise found in institutions that do not primarily address this region.

In the normative domain as well, nesting under broader international institutions is important. We have noted that the various guidance documents prepared under the Arctic Council highlight norms elaborated within global organizations or under regional marine pollution regimes. Similarly, the working groups on conservation under the BEAR and the Arctic Council draw upon the global CBD process for both normative and implementation guidance. Although the CBD itself is rather general in its obligations, the 'CBD machinery' works well for developing advice and guidance on implementation of the various obligations through extensive work programmes for different biomes, and through guidelines and recommendations, including on protected areas.

There are also instances when the Arctic institutions have acted as catalysts, seeking to influence normative developments under other institutions. Of particular interest in this respect is the relationship between the Arctic Council and the adoption of stronger rules on POPs, first in the Århus POPs Protocol under CLRTAP and then in the global Stockholm Convention.[40] Following the adoption of AEPS, delegates from the eight Arctic states expressed their concern about the Arctic health effects of POPs before the Executive Body of CLR-TAP. Given the political weight of this group of states, this is likely to have contributed to a subsequent strengthening of the mandate given to the task force that examined the basis for a POPs protocol (Reiersen *et al.* 2003: 61). Moreover, the period of producing the first AMAP Assessment overlapped with the negotiation of the POPs Protocol, and the fact that the AMAP chair was also co-chair of the Task Force on POPs implied that Arctic findings were continuously fed into the process.[41] It seems likely that those findings added to the perception that there existed a firm scientific basis for action. On the other hand, the impacts of Arctic Council activities on the adoption or contents of the Århus Protocol should not be overstated. As noted, knowledge about extraordinarily high exposure to POPs among some Arctic residents preceded AMAP. Selin (2000; 2003), in his in-depth study of the POPs Protocol, highlights research and monitoring activities by certain Arctic states – especially Canada and Sweden – rather than AMAP or other Arctic Council activities. Indeed, Selin (2000: 133) reports that the early emphasis of the POPs Task Force on the adverse impacts on Arctic and sub-Arctic residents was seen as a tactical error that limited broader political support for a new protocol; this perception induced a broadening of impacts studies to include the entire CLRTAP region.

With the global Stockholm Convention on POPs, adopted three years after the regional accord, Inuit organizations played an active role, which they had done only towards the end of the Århus Protocol process (Fenge 2003). Particular attention was given to engaging US support for an ambitious instrument, and the US chairmanship of the Arctic Council in the period 1998–2000 provided a platform for exerting Alaskan influence on those who defined US positions in the global POPs negotiations (Huntington and Sparck 2003: 221–2). The 2000 Barrow Declaration, in which Arctic ministers agreed to 'strengthen efforts to finalize a comprehensive and verifiable convention at the last session of the Intergovernmental Negotiations Committee', indicated a more positive US attitude to an ambitious instrument, and the USA was among the signatories to the Stockholm Convention. According to the Conference Chair, representatives of indigenous and environmental organizations influenced the multilateral negotiations in a productive way and also ensured sustained publicity on the process (Buccini 2003: 250). The Arctic dimension of the POPs problem was symbolized by a carving of an Inuit woman that was placed on the Chair's table throughout the multilateral negotiations, but caution is again required in trying to assess the impacts of the Arctic Council on the negotiated outcome. It seems safe to observe that findings generated under AMAP formed part of the scientific basis that substantiated the need for stricter regulations; that participation by Inuit organizations in the Arctic Council strengthened their awareness and knowledge about international POPs politics and improved their access to, and general relations with, Canadian foreign policy makers; and that Arctic Council meetings held in the USA provided platforms for actors (including in Alaska) who favoured an ambitious global instrument to exert pressure on US foreign policy makers.

Another type of catalytic activity involves using Arctic institutions to encourage regional states to join or participate more extensively in spatially broader environmental instruments. Among the legal instruments identified under Arctic Council working groups as in need of broader participation by Arctic states, the Law of the Sea Convention has subsequently been ratified by Russia and Denmark. Given the comprehensive nature of that instrument, it seems unlikely, however, that the signals provided under the Arctic Council were significant drivers behind those decisions; and it should be borne in mind that the USA has remained outside the Convention. International support to enhance Russia's facilities for treating liquid radioactive waste has enabled that state to join the ban on dumping of low-level radioactive waste (IMO 2005) – but those projects were not financed or organized under the BEAR or the Arctic Council. Similarly, despite continuous encouragement, also in several Ministerial Declarations, Russia has remained outside the comprehensive and ambitious OSPAR Convention on marine pollution in the Northeast Atlantic; neither has it joined the Århus POPs Protocol, or ratified the Stockholm Convention yet.[42] Several other Arctic states too have failed to ratify the two POPs instruments.

Also capacity enhancement efforts under Arctic institutions involve institutional interplay. Arctic Council activities to reduce discharges of hazardous substances bound for the Arctic are nested within broader normative endeavours and programme activities – and also trigger external financial support for projects. PAME's Regional Plan of Action to Eliminate Pollution from Land-Based Activities is explicitly linked to UNEP's corresponding global endeavour.[43] Under this programme, PAME has been instrumental in attracting financial support for the Russian National Plan of Action from international financing institutions, including several projects approved by the Global Environment

Facility (GEF) and NEFCO. Dissatisfaction among some Arctic states about the balance between funds allocated for planning and preparatory purposes on the one hand and actual removal of environmental threats on the other, has constrained fund-raising for these projects and achievements so far are very limited.[44] The UNEP link is also strong with respect to ACAP: for some time, UNEP-Chemicals provided secretarial services for the ACAP project on obsolete pesticides and is also involved in projects that address mercury, dioxins and furans.

Political mobilization

Indigenous peoples' organizations, research communities and regional-level authorities are arguably those most directly empowered by the environmental collaboration that has been developed under the Arctic Council and the BEAR. Effective involvement of Arctic indigenous peoples is a salient element in the founding documents of Arctic institutions. In the Rovaniemi Declaration, the Arctic states emphasized their 'responsibility to protect and preserve the Arctic environment. . . recognizing the special relationship of the indigenous peoples and local populations to the Arctic and their unique contribution to the protection of the Arctic environment'. This was reinforced by the 1996 Ottawa Declaration, which elevated the status of some indigenous organization to that of Permanent Participants. In the BEAR cooperation too, indigenous peoples were represented in the Regional Council, and indigenous organizations have taken part in its working groups, including that on the environment. The impacts of the BEAR and the Arctic Council on indigenous participation in, and influence on, Arctic decision making is discussed in greater depth by Wilson and Øverland in Chapter 3. The extent of actual involvement differs across issues, but it seems that the Arctic Council in particular has provided indigenous organizations with information and networks that have helped to amplify their voice in national and international environmental fora on matters of particular importance to them, especially hazardous substances. Additionally, indigenous organizations have had a say on data collection, parameter choice and research foci, for instance under AMAP and the ACIA.[45]

Scientists are another category of actors that Arctic institutions have enabled to build denser and more extensive trans-national networks. The International Arctic Science Committee emerged prior to the Arctic Council and the BEAR, but the emphasis of the latter two on environmental monitoring and collaborative research has generated funds that have made possible the development of stable contacts among a great many scientists and research institutions. In part, this has taken place through participation in BEAR or Arctic Council working groups or in the expert groups established to produce specific reports. More than 300 scientists were involved in the preparation of the first AMAP Assessment Report, many of whom also worked together on the second, and similar numbers contributed to the ACIA reports. For those purposes, funds have been allocated by Western states to enable participation by financially troubled research institutions in Russia. The new institutions have also strengthened pre-existing researcher ties. Ever since the creation of BEAR, for instance, Norway's Institute of Marine Research has been able to use substantial Foreign Ministry funds set aside for Barents cooperation to reinforce collaboration with its counterpart in Murmansk, the Knipovich Polar Marine Research Institute of Marine Fisheries and Oceanography (PINRO). The latter is central to monitoring and assessment of fish stocks and environmental conditions in the Russian part of the Barents Sea. An express motive for that support is to ensure personnel continuity and data quality on the Russian side in areas particularly critical to joint assessment programmes.

The new Arctic institutions have also strengthened networks among environmental civil servants and provided access to sources of information and environmental decision makers in other countries that would otherwise have been unavailable. This is particularly true for the province level of collaboration. In Norway, for instance, it was decided that all Local Agenda 21 projects in the three northernmost counties should be carried out in collaboration with twin towns in Northwestern Russia. Contacts have been established between Russia and the Nordic states, but, as in the health sector, cross-border environmental activities have also served to link the Northwest Russian federal subjects closer to each other.[46]

As noted above, at the intergovernmental level, bilateral environmental cooperation preceded the Arctic Council and the BEAR, and probably would have evolved even without them. On the other hand – and to an extent that varies from one issue area to another – the Arctic institutions have added a multilateral layer that shapes that nature of networks. With respect to Arctic conservation issues, for instance, participants in Arctic working groups have sensed a change over the years, in that representatives from different countries increasingly consider each other as partners in a teamwork endeavour to promote biodiversity, and not so much as defenders of national short-term interests. The value of such teams is enhanced by the fact that scientists, civil servants and representatives of environmental organizations and indigenous groups are working together and creating ownership among the various stakeholders. Both the BEAR and the Arctic Council have meant a central role for Arctic ministries of foreign affairs in regional networks on sustainable development matters. Some of those engaged in Arctic Council activities look back on the AEPS era, when environmental ministries were in the driver's seat, with a certain nostalgia (Haavisto 2001). On the other hand, access to foreign affairs officials is often necessary when the problems addressed span several sectors or require international responses beyond those of the Arctic eight.

When the BEAR cooperation was set up, Western participants were aware that Russian civil society had been quite poorly developed, also in the environmental area (Stokke 1994). Semi-official organizations for nature protection had thrived throughout the Soviet era, but a critical environmental movement independently of state authorities – so-called 'informals' – did not emerge until the mid-1980s (DeBardeleben 1992). Strengthening ties between Nordic environmental organizations and those in Russia became a priority area and some Western funds were allocated for this purpose. A certain amount of mobilization has occurred among Russian environmental activists as a result, but the overall effect has been ambiguous: Russian environmental civil society groups remain weak and progress is impeded by the suspicion that surrounds organizations that receive funding from abroad.[47] On the Nordic side, in contrast, several 'green' organizations have clearly improved their standing and influence in domestic debates as a consequence of their trans-national connections. This is particularly true for Norway's Bellona, which has played a highly visible role in compiling and disseminating information about Soviet and Russian nuclear activity that had not been available in the official bilateral channels.

In the Arctic Council, some environmental organizations have taken an active role in projects and at meetings, especially the World Wide Fund for Nature (WWF) and the Advisory Committee on Protection of the Sea (ACOPS). The latter has participated in PAME activities related to the Russian National Plan of Action, a role that has been subject to some controversy among Arctic states. The WWF too has taken part in Arctic Council projects and produces the *Arctic Bulletin*, a newsletter that pays considerable attention to Arctic Council and BEAR activities. It is not clear whether these organizations have improved their public visibility or ability to influence important decisions as a consequence of their engagement, however. Both are large, well-established organizations whose Arctic projects represent only a portion of their activity. Not surprisingly, some environmental-group participants in Arctic Council work have expressed frustration at too little progress, too few resources allocated to environmental protection, and too many opportunities lost in the Arctic cooperation. On the other hand, there are no indications that the civil society organizations involved have any feelings of being hostages to the cooperation – committed by participation to support policies they oppose – as has occurred in some other cases of environmental diplomacy with trans-national involvement.

Region building

Nurtured by the new Arctic institutions, networks among governments, provincial authorities, indigenous organizations, scientists, and civil society organizations enable more face-to-face interaction and communication among regional actors than would otherwise have occurred. These various actors come together on a regular basis in arenas that have now achieved a certain track record and momentum – making it important to take seriously the expectations that one will propose and implement tangible programmes, and respond to the proposals of others. In turn, this encourages those involved to consider planned activities or distinct challenges they face through an Arctic, or Euro-Arctic prism: to some extent at least, Arctic institutions support a Northern framing of environmental problems and initiatives. It has become necessary for governments and even province-level authorities to develop an 'Arctic policy'. On the ground, working groups under the Arctic Council and BEAR have created networks where people with expertise on and competence in monitoring, marine pollution, emergency response, or nature conservation are given shared tasks – a practice conducive to the development of more lasting connections. Experience has shown that such networks are extremely valuable in promoting international work, and that individuals who have worked together in such networks tend to stay in touch in future positions where similar issues are dealt with.

We noted above that much of the environmental cooperation between Russia and the Nordic states has occurred outside the BEAR institutions. Bilaterally financed projects are reported as 'Barents projects', and large multilateral projects are financed not by the national Barents secretariats, but largely by EU regional development programmes and also by the Arctic Council. Nevertheless, the BEAR working groups on the environment have had a coordinating role in a multitude of activities, and most interestingly, a large number of projects seem to be perceived by many – both participants and observers – as 'Barents projects'. This is interesting from a region-building point of view, since it indicates that the Barents region has come to be seen as a 'natural' geographical, and perhaps also institutional, category. The BEAR institutions themselves have admittedly been relegated to a coordinating role, but even activities financed as bilateral projects or by the EU or various international organizations now appear as parts of the big Barents family of cross-border environmental project activity. Especially in the first years of BEAR collaboration, the environment also had a region-building effect in itself, with environmental rhetoric employed forcefully by BEAR players to convey an image of a single region with common concerns.

Regionality also has an external dimension, and there can be no doubt that Arctic Council activities in particular have enhanced the awareness and recognition among states, organizations and individuals outside the region of the Arctic dimension of many environmental problems. The AMAP Assessment Reports are important cases in point, not least with respect to hazardous substances and their impacts on human health in the context of international POPs politics, and this is true for the ACIA Report as well. As discussed by Hoel in Chapter 6, Arctic Council work on climate change is being fed into the global climate regime, and the Fourth Assessment Report under preparation by the Intergovernmental Panel on Climate Change will include a separate chapter on the Arctic.

Conclusions

This chapter has narrowed in on four items that weigh heavily on the Arctic environmental agenda: improving monitoring, reducing discharges of hazardous substances, enhancing nuclear safety, and protecting biodiversity. Having considered programmatic and normative activities under the Arctic Council and the Barents Euro-Arctic Region (BEAR), we would like to make six points about their impacts on environmental problem solving, political empowerment, and regional connectedness.

First, environmental monitoring has emerged as a specialization of the Arctic Council and is an area of activity where this institution has made a significant difference, primarily by providing means for coordination of national and international programmes on the Arctic and by encouraging states to allocate more resources for this purpose. Monitoring activities are also coordinated within the Barents region, but largely at the bilateral level and not as a consequence of BEAR activities. Although awareness of the adverse effects of hazardous substances on Arctic animals and human beings preceded the Arctic Monitoring and Assessment Programme, the latter has contributed to the substantiation of sources, pathways and effects, sometimes in collaboration with other working groups under the Arctic Council.

Second, this information has been fed into broader international efforts to regulate discharges of persistent organic pollutants and heavy metals, and has contributed, albeit hardly decisively, to strengthening the position of those who favoured more ambitious regulation. Such interplay with other institutions is necessary if Arctic institutions are to reduce regional pollution levels effectively, since many of the hazardous compounds that threaten Arctic ecosystems originate in other parts of the world. Institutional interplay is also important for efficient monitoring and for generating the resources needed to reduce environmental threats in the area. Most of the monitoring, technology transfer and construction of storage and treatment facilities to improve nuclear safety in Northwestern Russia has been organized and financed by institutions other than those examined here, largely on a bilateral basis or drawing upon EU or US funds.

Third, the Arctic Council in particular has invested considerable energy in developing guidelines on the safe conduct of Arctic operations, especially with respect to oil, gas and shipping activities, and for certain specific conservation issues. None of those are legally binding, however, and to a large extent they apply norms that originate in broader international processes. Since there are no structures or procedures for systematic review of whether those soft law instruments are implemented by Arctic states or operators, it is difficult to judge their influence, but there is little to indicate that such influence is high.

Fourth, the BEAR and the Arctic Council have worked to identify environmental 'hot spots' in Russia and sought to extract funds from Arctic states or from global or regional environmental financing institutions in order to compile, store and destroy selected contaminants, like PCB and pesticides, and to identify and remove sources of contaminants in drinking-water basins. While several of these projects have run into difficulties, either with respect to financing or with site selection, they stand as promising examples of regional capacity enhancement. If successful, they will show that Arctic institutions can do more than merely pinpoint environmental threats: they can also generate the resources needed to address them.

Fifth, networks generated and maintained by the Arctic Council and the BEAR have had some mobilizing, or empowering, effect on indigenous peoples' organizations, environmental researchers and civil servants at the regional level of governance. Some environmental organizations have also improved their access to Arctic decision makers as a consequence of those institutions, but the role of environmental civil society organizations remains weak, especially in Russia.

Finally, the networks that have emerged within the environmental sector of the Arctic Council and the BEAR encourage participants to view the environmental challenges faced in the region within a Northern frame. Widely disseminated assessments of the state of the Northern environment have nurtured the discursive distinctiveness of the Arctic region, among regional players and outsiders alike. Again, such a framing is limited by the circumstance that some of the most pressing Arctic pollution issues originate in industrial activities further south – and solving even some of those that relate to regional activities, such as nuclear safety, will require participation from outside the Arctic.

Notes

1 Very helpful comments are appreciated from Lars-Otto Reiersen, David Scrivener, Oran Young and our fellow contributors to this book.
2 Declaration on the Establishment of the Arctic Council (1996), Art. 1.a.
3 See Declaration on the Establishment of the Arctic Council (1996), Art. 1.b; and Declaration on the Protection of the Arctic Environment (1991).
4 Declaration of the Conference on Foreign Ministers of the Baltic Sea States (1992).
5 Declaration on Cooperation in the Barents Euro-Arctic Region (1993). The Rio Declaration contains a set of general environmental principles and Agenda 21 is a detailed description of selected global environmental challenges and recommendations on ways to address them. Both were adopted at the 1992 UN Conference on Environment and Development, held in Rio de Janeiro in 1992.
6 The 'first international environmental wave' had come in the 1970s, symbolized by the UN Conference on the Human Environment, held in Stockholm in 1972.
7 On the distinction between normative and programmatic activities under international regimes, see Young (1999: 26–35).
8 On the impacts of climate change on living conditions for polar bears, see Chapter 6 by Hoel; on the threats associated with higher catch quotas in Canada and the lack of quotas in Greenland, see Wiig (2005).
9 The 1997 Alta Ministerial, held between the founding Ottawa Ministerial and the first Arctic Council Ministerial, endorsed the publication of the first AMAP (Arctic Monitoring and Assessment Programme) Assessment Report. On AMAP, see below.
10 Alta Ministerial Declaration.
11 The BEAC Environmental Action Plan was adopted in Bodø, 1994.
12 See Chapter 1; the formation of the AEPS is described and analysed by Young (1998).
13 On the transition from the AEPS to the Arctic Council, see Scrivener (1996) and Young (1996).

14 See AMAP (1997; 2002); the second report updates the first with respect to POPs, heavy metals, radioactivity, human health and changing pathways.

15 The four-year Arctic Climate Impact Assessment, examined by Hoel in Chapter 6, was a joint undertaking involving AMAP, CAFF and the International Arctic Science Committee. The Oil and Gas Assessment is discussed by Offerdal in Chapter 7.

16 The assessment work plan is available at the PAME website (www.pame. is); Canada, Finland and the USA are lead countries. A PAME 'snapshot analysis', produced under the auspices of the Norwegian Maritime Directorate, was published in 2000.

17 See EPPR (1998).

18 See for instance PAME (1996).

19 See the discussion by Offerdal in Chapter 7.

20 See, respectively EPPR (1998) and PAME (2004).

21 On the limited progress in this line of PAME activity, see below.

22 Hot spots' are sites or geographic areas marked by particular environmental disturbance due to man-made activities The identification of such hot spots has occurred in collaboration with the Nordic Environment Finance Corporation and AMAP (AMAP 2003).

23 Those EU funds were allocated from its Interreg programme. In 1996, the EU Commission had established two new Interreg programmes, Barents and North Calotte, based, respectively, in Luleå and Rovaniemi. By providing fresh funding, Norway became an associate member of both.

24 Bente Christiansen, cited in Finland, Ministry of the Environment (2003: 47).

25 On a similar tendency in the health sector, see Chapter 4 by Rowe and Hønneland.

26 See for instance AMAP (2006), which highlights possibilities for collaboration on projects that address Russian hot spots or follow-up on the Oil and Gas Assessment.

27 An AMAP project finalized in 2001 established an air monitoring station in Anderma (in Nenets Autonomous Area) for sampling and analysis of POPs and mercury.

28 On this and other Arctic Council activities, see (www.arctic-council.org/ en/main/infopage/5#a4).

29 See AMAP (2006). The Thematic Data Centre for atmospheric data is located at the Norwegian Institute for Air Research; the one for marine data at the International Council for the Exploration of the Sea, Denmark; the centre for freshwater and terrestrial data at the University of Alaska-Fairbanks, USA; the one for radioactivity at the Norwegian Radiation Protection Agency; and the compilation of health- and climate-related data is coordinated by the AMAP Secretariat, based in Norway.

30 See (www.amap.no). The 'overview' report of the Arctic Climate Impact Assessment has been translated into the Russian, French, German, Dutch, Norwegian and Saami languages (AMAP 2005: 2–3).

31 See ACAP (2006: 3); the three-year plan and its relationship to earlier phases of the project is sketched in ACAP (2002).

32 Estimated costs of these destruction projects were estimated at around USD 5 million and USD 8 million in equipment; see ACAP (2003), App. I.

33 See ACAP (2003; 2005; 2006); the technology envisaged is the same that is to be used for PCB destruction.
34 The project plan is sketched in ACAP (2005).
35 See (eppr.arctic-council.org/fldguide/index.html).
36 The AMEC collaboration was established by Norway, Russia and the USA in 1996. The UK joined in 2003. AMEC has provided funding for dismantling of nuclear submarines and handling of radioactive waste in Northwestern Russia; Norwegian funding again comes from the Plan of Action. A few nuclear safety projects have received financing from the Barents Secretariat in Kirkenes, but compared to the Plan of Action this source of financing is quite insignificant.
37 The agreement was signed by Belgium, Denmark, Finland, France, Germany, the Netherlands, Norway, Russia, Sweden, the UK, the USA, the EU and EURATOM.
38 On the notion of institutional interplay, see Chapters 2 and 8 in this book.
39 Fish stock assessment is confined to the Northeast Atlantic.
40 On this matter, see also the discussion by Wilson and Øverland in Chapter 3.
41 See Reiersen *et al.* (2003: 68); the chair/co-chair was David Stone of Canada.
42 With respect to the Århus POPs Protocol, Russia has not even signed. For updated status of this instrument see (www.unece.org) and for the Stockholm Convention (www.pops.int).
43 See UNEP (1995); information about ongoing programme activities is available at (www.unep.org).
44 Compare PAME (2002), which reports GEF approval of funding for projects, with PAME (2003, 2006), which report some meeting activity.
45 In ACIA, several Permanent Participants were involved in the Assessment Steering Committee, and programme work was tailored to take account of indigenous experience of and knowledge on climate change: see Chapter 6 by Hoel.
46 On such impacts of health-related programmes, see Chapter 4 by Rowe and Hønneland.
47 On the financial dependency of some Russian civil society organizations on Western partners, see Henderson (2003). Skedsmo (2005) offers an anthropological study of 'everyday practices' in an environmental organization in Murmansk, initiated by a Norwegian 'sister organization'. One of his findings is that the prospect of accumulating individual social capital is a main motivation for joining a Western-sponsored civil society organization. Although altruism is not absent, members were found to be oriented more towards individual career building (e.g. aiming to progress to other types of international cooperation, such as business) than towards building civil society in Russia.

References

ACAP (Arctic Council Action Plan to Eliminate Pollution in the Arctic) (2002) 'ACAP progress report to Senior Arctic Officials' (Saariselka, Finland, 7–8 October 2002). www.arctic-council.org.

ACAP (Arctic Council Action Plan to Eliminate Pollution in the Arctic) (2003) 'ACAP progress report to Senior Arctic Officials' (Svartsengi, Iceland, 23–24 October 2006). www.arctic-council.org.

ACAP (Arctic Council Action Plan to Eliminate Pollution in the Arctic) (2005) 'ACAP progress report to Senior Arctic Officials' (Yakutsk, Russia, 6–7 April 2005). www.arctic-council.org.

ACAP (Arctic Council Action Plan to Eliminate Pollution in the Arctic) (2006) 'ACAP progress report to Senior Arctic Officials' (Syktyvkar, Russia, 26–27 April 2006). www.arctic-council.org.

AMAP (Arctic Monitoring and Assessment Programme) (1997) *Arctic Pollution Issues: A State of the Arctic Environment Report*, Oslo: Arctic Monitoring and Assessment Programme. www.amap.no.

AMAP (Arctic Monitoring and Assessment Programme) (2002) *Arctic Pollution 2002*, Oslo: Arctic Monitoring and Assessment Programme. www.amap.no.

AMAP (Arctic Monitoring and Assessment Programme) (2003) *Updating of Environmental 'Hot Spots' List in the Russian Part of the Barents Region*, Oslo: Arctic Monitoring and Assessment Programme.

AMAP (Arctic Monitoring and Assessment Programme) (2004) 'AMAP strategy 2004+'. Oslo: Arctic Monitoring and Assessment Programme. www.amap.no.

AMAP (Arctic Monitoring and Assessment Programme) (2006) 'Progress report from the AMAP Board to the SAO meeting' (Syktyvkar, Russia, 26–27 April 2006). www.arctic-council.org.

Archer, Clive and David Scrivener (2000) 'International cooperation in the Arctic environment' in M. Muttall and T. V. Callaghan (eds.), *The Arctic: Environment, People, Policy*, Amsterdam: Harwood Academic, pp. 601–20.

BEAC (Barents Euro-Arctic Council) (1994) 'Environmental Action Programme'. www.barentsinfo.fi/wge_docs/Action_prog_1994.pdf.

Bloom, Evan T. (1999) 'Current developments – establishment of the Arctic Council', *American Journal of International Law*, 93: 712–22.

Buccini, John Anthony (2003) 'The long and winding road to Stockholm: the view from the Chair' in D. L. Downie and T. Fenge (eds.), *Northern Lights Against POPs: Combatting Toxic Threats in the Arctic*, Montreal: McGill–Queen's University Press, pp. 224–56.

DeBardeleben, Joan (1992) 'The new politics in the USSR: the case of the environment' in J.M. Stewart (ed.), *The Soviet Environment: Problems, Policies and Politics*, Cambridge: Cambridge University Press, pp. 64–87.

Declaration from the Conference of Foreign Ministers of the Baltic Sea States (Copenhagen, 6 March 1992). www.cbss.st.

Declaration on Cooperation in the Barents Euro-Arctic Region, Conference of Foreign Ministers (Kirkenes, 11 January 1993). www.beac.st.

Declaration on the Establishment of the Arctic Council (Ottawa, 19 September 1996). www.arctic-council.org.

Declaration on the Protection of the Arctic Environment (Rovaniemi, 14 June 1991). www.arctic-council.org.

Dewailly, Eric and Christopher Furgal (2003) 'POPs, the environment, and public health' in D. L. Downie and T. Fenge (eds.), *Northern Lights Against POPs: CombattingToxic Threats in the Arctic*, Montreal: McGill–Queen's University Press, pp. 3–21.

EPPR (Emergency Prevention, Preparedness and Response) (1998) 'Environmental risk analysis of Arctic activities.' www.eppr.arctic-council.org.

Fenge, Terry (2003) 'POPs and Inuit: influencing the global agenda' in D. L. Downie and T. Fenge (eds.), *Northern Lights Against POPs: Combatting Toxic Threats in the Arctic*, Montreal: McGill–Queen's University Press, pp. 192–213.

Finland, Ministry of the Environment (2003) *Environmental Cooperation in the Barents Region: 10-Year Review*, Helsinki: Finnish Ministry of the Environment.

Haavisto, Pekka (2001) *Review of the Arctic Council Structures*, Helsinki: Finnish Institute of International Affairs.

Henderson, Sarah L. (2002) 'Selling civil society: Western aid and the non-governmental organization sector in Russia', *Comparative Political Studies*, 35: 139–67.

Hønneland, Geir (2003) *Russia and the West: Environmental Cooperation and Conflict*, London: Routledge.

Hønneland, Geir and Arild Moe (2000) 'Evaluation of the Norwegian Plan of Action for Nuclear Safety: priorities, organisation, implementation', *Evaluation Report 7/2000*, Oslo: Norwegian Ministry of Foreign Affairs.

Huntington, Henry P. and Michelle Sparck (2003) 'POPs in Alaska: engaging the USA' in D. L. Downie and T. Fenge (eds.), *Northern Lights Against POPs: Combatting Toxic Threats in the Arctic*, Montreal: McGill–Queen's University Press, pp. 214–23.

IMO (International Maritime Organization) (2005) 'Russian Federation accepts ban on dumping of radioactive wastes under 1972 London Convention', *IMO Briefing 26/2005*.

Owens, Edward H., and Gary A. Sergy (2004) *The Arctic SCAT Manual: A Field Guide to the Documentation of Oiled Shorelines in Arctic Environments*, Edmonton: Environment Canada.

PAME (Working Group on Protection of the Marine Environment) (1996) 'Report to the third ministerial conference on the protection of the Arctic environment' (Inuvik, Canada, 20–21 March 1996). www.arctic.council.org.

PAME (Working Group on Protection of the Marine Environment) (2002) 'PAME report to Senior Arctic Officials' (Oulo, Finland, 15–16 May 2002). www.arctic-council.org.

PAME (Working Group on Protection of the Marine Environment) (2003) 'PAME report to Senior Arctic Officials' (Reykjavik, Iceland, 9–10 April 2003). www.arctic-council.org.

PAME (Working Group on Protection of the Marine Environment) (2004)

'Guidelines for Transfers of Refined Oil and Oil Products in Arctic Waters'. www.pame.is.

PAME (Working Group on Protection of the Marine Environment) (2006) 'PAME report to Senior Arctic Officials' (Syktyvkar, Russia, 26–27 April 2006). www.arctic-council.org.

Reiersen, Lars-Otto, Simon Wilson and Vitaly Kimstach (2003) 'Circumpolar perspectives on persistent organic pollutants: the Arctic Monitoring and Assessment Programme' in D. L. Downie and T. Fenge (eds.), *Northern Lights Against POPs: Combatting Toxic Threats in the Arctic*, Montreal: McGill–Queen's University Press, pp. 60–86 .

Scrivener, David (1996) 'Environmental cooperation in the Arctic: from strategy to council', *Security Policy Library* 1, Oslo: The Norwegian Atlantic Committee.

Selin, Henrik (2000) *Towards International Chemical Safety: Taking Action on Persistent Organic Pollutants (POPs)*, Linköping: Department of Water and Environmental Studies, Linköping University.

Selin, Henrik (2003) 'The UNECE CLRTAP POPs Protocol' in D. L. Downie and T. Fenge (eds.), *Northern Lights Against POPs: CombattingToxic Threats in the Arctic*, Montreal: McGill–Queen's University Press, pp. 111–32.

Skedsmo, Pål (2005) 'Doing good in Murmansk? Civil society, ideology and everyday practices in a Russian environmental NGO', *FNI Report* 14/2005, Lysaker: The Fridtjof Nansen Institute.

Stokke, Olav Schram (1990) 'The Northern environment: is cooperation coming?' *Annals of the American Academy for Political and Social Science* 512: 58–69.

Stokke, Olav Schram (1992) 'Arctic environmental cooperation after Rovaniemi: what now?' in L. Lyck (ed.), *Nordic Arctic Research on Contemporary Arctic Problems*, Aalborg: Aalborg University Press, pp. 223–36.

Stokke, Olav Schram (1994) 'Environmental cooperation as a driving force in the Barents Region' in O. S. Stokke and O. Tunander (eds.), *The Barents Region: Cooperation in Arctic Europe*, London: SAGE, pp. 145–159.

Stokke, Olav Schram (1998) 'Nuclear dumping in Arctic seas: Russian implementation of the London Convention' in D. G. Victor, K. Raustiala and E. B. Skolnikoff (eds.), *The Implementation and Effectiveness of International Environmental Commitments: Theory and Practice*, Cambridge, MA: MIT Press, pp. 475–517.

Stokke, Olav Schram (2000) 'Sub-regional cooperation and protection of the Arctic marine environment: the Barents Sea' in D. Vidas (ed.), *Protecting the Polar Marine Environment: Law and Policy for Pollution Prevention*, Cambridge: Cambridge University Press, pp. 124–48.

UNEP (United Nations Environment Programme) (1995) *Global Programme of Action for the Protection of the Marine Environment from Land-Based Activities*, UNEP (OCA)/LBA/IG.2/7. www.unep.org.

Wiig, Ø. (2005) 'Isbjørnen truet av utryddelse' ('Polar bear threatened with extinction'). www.aftenposten.no/meninger/kronikker/article1121817.ece.

Young, Oran R. (1996) *The Arctic Council: Marking a New Era in Internation.*
Relations, New York: The Twentieth Century Fund.

Young, Oran R. (1998) *Creating Regimes: Arctic Accords and International*
Governance, Ithaca, NY: Cornell University Press.

Young, Oran R. (1999) *Governance in World Affairs*, Ithaca, NY: Cornell University Press.

mate change

Håkon Hoel

Introduction

The last mammal brought about by evolution, the polar bear, is facing extinction. Global warming is about to change the habitat of polar bears in a way that threatens their very survival as a species (Klein 2005). As a harbinger of the changes that climate change may bring to Arctic ecosystems, the polar bears are but one example in a catalogue of challenges confronting the Arctic nations. The Arctic region plays a critical role in the global climate system; changes to the climate there have global ramifications. Moreover, the effects of climate change are particularly pronounced in the Arctic (ACIA 2004, 2005). In recent decades, average temperatures have risen at rates almost twice that of the rest of the world, and are expected to increase substantially more than the global average in the course of this century.

The possible consequences of climate change have spurred a major institutional response at the global level. Since 1988, the Intergovern-mental Panel on Climate Change (IPCC) has produced assessments of the status of science in this realm (IPCC 1990, 1996, 2001). In 1992, the UN Framework Convention on Climate Change (UNFCCC)[1] was adopted, attracting virtually global membership. It was followed up by the Kyoto Protocol[2] in 1997, which contains specific objectives and measures for reductions in emissions of climate gases and timetables for achieving them.

This chapter examines how the Arctic nations are responding to the issue of climate change and discusses the nature and potential impacts of those responses. Following a brief overview of the climate change issue, we look into the institutional responses, as played out in the Arctic Council and in the eight Arctic countries.[3] Fundamentally, their collective response has been one of studying the problem, rather than taking mitigating or adaptive action. The institutional developments are accounted for in terms of their impacts, understood as effectiveness, mobilization of actors, and contribution to regionalization (cf. Chapter 2 of this volume).

The climate problem: warmer, wetter and wilder

Throughout recorded history, a distinctive feature of the global climate has been its variability. Over the past century, global temperatures appear to have risen at rates of change unprecedented for millennia (Dessler and Parson 2006). In the Northern Hemisphere, the last century witnessed more pronounced warming than any other period during the past 1,000 years (Osborne and Briffa 2006).

Attributed to the burning of fossil fuels, 'greenhouse gases' denote a phenomenon where the emission of certain gases (CO_2 in particular) into the atmosphere traps the outgoing radiation of heat from the Earth.[4] The result is that the temperature in the atmosphere increases, in a process that feeds on itself by reducing snow and ice cover. The surface that admits incoming solar radiation is increased, and significantly reduces the Earth's albedo. Since the polar areas have vast areas that are covered by snow and ice, these regions are likely to be more affected by climate change than areas at lower latitudes.

Associated with rising temperatures are various changes in the biological and physical environment. Climate scenarios indicate such highly significant changes at the global level (IPCC 2001) as:

- continued increase in temperatures throughout this century (+1.4 to +5.8°C);
- a rise in sea level in the 10–90 cm range;
- disturbances in oceanic circulation patterns;
- shifts in vegetation patterns and in the geographic distribution of animal species;
- changes in precipitation patterns;
- increased frequency of storms.

In the Arctic, temperatures rise faster than global average. The most recent assessment (ACIA 2005) indicates that atmospheric temperatures in the Arctic as a whole are increasing at about twice the rate of the global average, at 4–7°C in the course of this century. The Arctic region is therefore set to experience a series of climate change related perturbations that is likely to affect the environment, people and societies in the region in numerous ways (ACIA 2004, 2005). Increased temperatures will bring shorter and warmer winters, and snow and ice cover will decrease. Vegetation zones will shift, with the treeline moving northwards. The productivity of ecosystems will change, as will the geographical distribution and diversity of species. Sea-ice extent and thickness will decrease – by the late twenty-first century, the central Arctic Ocean may be ice-free in summer. The area covered by permafrost is likely to decrease considerably.

A major aspect of climate change in the Arctic region is variation: while some areas appear to experience severe change (e.g. Alaska), others seem less affected (Northeast Atlantic). Moreover, the Arctic region is extremely diverse in terms of climate: due to the moderating influence of the Atlantic current, the Scandinavian countries and Iceland have a relatively temperate climate compared to other regions at the same latitude. North Norway, for instance, is on average 6–10°C warmer than areas at the same latitude in North America.

Institutional response: cold, dry and timid

The institutional response to the climate challenge has been a global regime, and should be considered in light of developments in the regime based on the 1992 UN Framework Convention on Climate Change (UNFCCC). The Arctic nations, except the USA (see below), are parties to the regulatory part of the regime, and work through this to confront the challenges posed by global warming.

In examining institutional responses to any given problem, it can be fruitful to distinguish between responses that are geared at enhancing knowledge about the problem, and responses that address the regulation of actor behaviour.

The global regime

Scope, membership, functions

The 1992 Framework Convention (in force since 1994) structures the interaction among states in the field of climate change at the global level. As the problem is global, so is the scope of the Convention. The UNFCCC introduces the principle of 'common but differentiated responsibilities', which charges industrialized countries ('Annex I countries') to limit their emissions of greenhouse gases (of which CO_2 is the most important) to 1990 levels. Less developed countries are exempted from the requirement to reduce emissions.

The obligations of Annex I countries in terms of specific reduction targets for emissions and timetables were negotiated in a separate treaty, the 1997 Kyoto Protocol. Here the Annex B countries, the most industrialized countries and some former East Bloc countries, agreed to reduce their emissions by 5.2 per cent on average in the period 2008–2012, relative to a 1990 baseline. Three mechanisms are devised for achieving efficient reductions of emissions on a global scale: joint implementation, the clean development mechanism, and emissions trading.[5] The Kyoto Protocol entered into force in 2005, following ratification by the Russian Federation late in 2004.[6] There are now 161 parties to the Protocol, accounting for 61.6 per cent of global emissions of climate gases. An important event in the development of this regime was the US decision in 2001 to withdraw from the work under the Kyoto Protocol.[7]

Work under the Convention and the Protocol relies on the scientific assessments of the Intergovernmental Panel on Climate Change (IPCC). The IPCC was established by the World Meteorological Organization (WMO) and the United Nations Environment Programme (UNEP) in 1988 (Bolin 1997). It is tasked with assessing the scientific, technical and socio-economic information on human-induced climate change, its impacts and options for adaptation and mitigation.[8] The IPCC works on the basis of peer-reviewed, published scientific literature, and is open to scientists from UN and WMO member nations. Participating scientists from more than 100 countries are nominated by governments and international organizations. The work is carried out in three working groups – on the climate system, vulnerability to climate change, and options for mitigation. Draft reports are subject to comprehensive peer reviews.

The main products of the IPCC are its Assessment Reports, accompanied by policy summaries for decision makers. As the IPCC is an independent body, the knowledge base of the global climate regime is produced by a body that is separate and distinct from the political wing of the climate regime.

Following the entry into force of the Kyoto Protocol, the sights are now set on the period after 2012, when Protocol's obligations expire. The 2005 Meeting of the Parties agreed to start talking about the issue in general terms, but did not agree on commitments.

Positions of the Arctic 8 on the climate issue

Arctic countries have varied in their attitude in the climate challenge, from the most ardent supporters of the global climate regime to its staunchest adversary, the United States. They have also varied with regard in their contributions to global emissions, with the USA at the one end of the scale (about one quarter of the global discharges of climate gases), and Iceland and Greenland on the other, with insignificant emissions.

In absolute figures, the USA is by far the largest emitter, with 6.9 billion metric tons of CO_2 equivalents in 2003, followed by Canada, Russia, Finland, Denmark, Sweden, Norway and Iceland (Table 6.1). Altogether the emissions of the Arctic 8 countries amount to some 8 billion tonnes of CO_2 equivalent, with the USA responsible for more than 85 per cent of this figure, or almost half the total emissions of climate gases by the Annex I countries in 2003.[9]

After 1990, Canada has had the biggest increase in emissions, with 24 per cent change, followed by Finland and the USA at 22 and 13 per cent respectively. Also Norway (9 per cent) and Denmark (7 per cent) have seen an increase in emissions. Sweden (–2 per cent), Iceland (–8 per cent), and Russia (–46 per cent) have decreasing emissions.

These figures say something about the difficulty the Arctic nations face in achieving the targets set by the Kyoto Protocol. Relative to the 1990 baseline, Canada, Finland and the USA in particular face severe challenges, but also Norway and Denmark will have to reduce their emissions or use the Kyoto mechanisms to achieve their targets.

In the past, the United States was a leading actor in international environmental policy collaboration, as witnessed for example by the negotiation of the global regime to protect stratospheric ozone (Benedick 1991). On the climate issue, however, the country has increasingly become a laggard. US opposition to the current global climate regime is due largely to the perceived burden on its economy from curbs on carbon emissions. Its position on the climate issue is that the question of explaining climate change has not been resolved, and that seeking to reduce climate gases by governmental regulations as the chief mitigation strategy is misplaced. The USA also takes issue with the fact that the Kyoto rules that stipulate reductions of discharges do not apply to developing countries (Victor 2004). It therefore withdrew from the work under the Kyoto Protocol in 2001.[10] However, the USA is also a major contributor to the work under the IPCC, and is a lead country in climate-related science. Both businesses and state authorities in the USA are increasingly concerned with the climate issue, and are preparing to adapt to carbon-reduction schemes.

Canada's energy consumption is increasing rapidly. The country is a supporter of the work under the Kyoto Protocol and is a recent chair of the Conference of the Parties. However, it is the second largest emitter

Table 6.1 Arctic country emissions in 2003 and change from 1990[a] (million tons of CO_2 equivalent)

	Total emissions 2003	Change from 1990 (%)
Canada	740,214	24
Denmark	75,485	7
Finland	85,559	22
Iceland	3,008	–8
Norway	54,779	9
Russia	14,2905	–46
Sweden	70,554	–2
USA	6,893,813	13
	8,066,317	19

a From Table 5, Doc. FCCC/SBI/2005/17. Note by the Secretariat to the Subsidiary Body for Implementation, 23rd session, Montreal 28 November–6 December 2005.

of climate gases, and has the highest rise of all in emissions since 1990 – a full 24 per cent increase.

As members of the European Union, Sweden, Finland, and Denmark play by EU rules on the climate issue. The EU has been the major driving force behind the Kyoto Protocol and its entry into force. While the legal competence of the Community to regulate member-state behaviour on the issue is less than straightforward,[11] it has nevertheless assumed a leading role in this regard. To meet its obligations, the EU established the first international emissions trading system in 2005, covering 12,000 enterprises in its 25 member states (Vogler and Bretherton 2006:6).[12] The EU has also proposed a reduction in climate-gas emissions for developed countries in the order of 15–30 per cent relative to 1990 levels by 2020, and 60–80 per cent by 2050 (Council 2005:10–11).[13]

The three EU Arctic nations – Denmark, Finland, and Sweden – differ considerably in terms of their achievement of Kyoto goals: Finland is in a difficult position, and Denmark has also seen its emissions growing. Sweden is the only Arctic country, along with Iceland (negligible emissions) and Russia, with emissions lower than the 1990 baseline.

The decline in industrial activity since the demise of the Soviet Union has brought Russia's emission levels far below its Kyoto targets. A relative latecomer to Protocol (2005), the Russian Federation made possible its entry into force.[14] As a major petroleum exporter, Russia is in a favourable position with its huge emissions deficit, and in a good position to exploit emission-trading arrangements.

Norway, like Russia, is a major petroleum-exporting country, but its steadily increasing emission levels have brought it about 10 per cent over 1990 levels. However, Norway has been generally supportive of the global climate regime.

Institutional responses in the Arctic

The Arctic Council

Arctic circumpolar cooperation on environmental protection was initiated in 1991 with the establishment of the Arctic Environmental Protection Strategy (AEPS) (see Stokke, Hønneland and Schei in Chapter 5 of this volume).[15] Original participants in the cooperation were the eight Arctic countries, as well as groups of indigenous peoples recognized as 'permanent participants' (see Wilson and Øverland, Chapter 3 of this volume). Several other countries and international organizations now participate as observers. The AEPS essentially focused on the collection and analysis of information on the status of the Arctic environment through a series of work programmes (Young 1998), with the Arctic Monitoring and Assessment Program (AMAP) the most relevant in our context.

The Arctic Council was established in 1996 by a declaration, rather than a treaty, and is therefore not geared towards legally binding regulation of state behaviour. The Arctic Council has largely continued the work of the AEPS, and the AMAP in particular has produced assessments of the environmental situation in the Artic that have commanded considerable attention. This has brought pollution issues to the table in forums where decisions about emission levels, timetables, and the like are actually made (AMAP 1998, 2004). Decisions of the Arctic Council are made by bi-annual ministerial meetings, in the form of non-binding declarations that give direction for future work under the Council. Day-today operations are taken care of by the countries' Senior Arctic Officials (SAOs), but work under the Council is dependent upon direct national financial contributions and willingness to act as lead country for projects.

The Arctic Climate Impact Assessment (ACIA)

The early work of the IPCC, as well as developments in climate-related science in general, soon indicated the importance of the Arctic region in this regard. The second AEPS ministerial in 1993 requested the AMAP to review the work of the global climate and ozone regimes and ensure that Arctic-specific issues were placed on the agenda of these bodies.

The climate issue had also come to the fore in the International Arctic Science Committee (IASC), a non-governmental body established in 1990 to initiate and coordinate Arctic science.[16] Since the early 1990s the IASC had addressed the issue of impacts of climate change, by starting up various research activities. In 1998 its executive committee proposed that the IASC should collaborate with the Arctic Council and the IPCC in developing a scientific assessment of the consequences of climate variability and change, and the effects of increased UV in the Arctic region.

In 1999, the AMAP, CAFF, IASC, IPCC and WCRP[17] collectively raised the idea of comprehensive assessment of the effects of climate change in the Arctic with governments and international bodies. An Assessment Steering Committee (ASC) was established, consisting of representatives from the Arctic Council programs and IASC. The Senior Arctic Officials responded by requesting an implementation plan for an Arctic climate impact assessment. Following workshops and consultations,[18] a formal implementation plan was submitted to the Senior Arctic Officials in September 2000, and was endorsed by the second ministerial meeting of the Arctic Council in Barrow later that autumn.[19] This constituted the formal start of the Arctic Climate Impact Assessment (ACIA) process. The USA was appointed lead country, in charge of driving the process and providing a secretariat for the project.

The ACIA was thereby formally established as an activity under the Arctic Council. The mandate provided in the declaration pointed to three tasks: the current scientific knowledge of the effects of climate change in the Arctic should be evaluated and synthesized, the possible consequences of climate change should be explored, and policy options addressed. The concept of an 'assessment' indicated that this was not to be an exercise in developing new knowledge or performing new science – the point was to take stock of existing knowledge of climate change in the Arctic region.

Organization and work of the ACIA

The implementation plan of the ACIA spells out the rationale for the project, its goals, potential benefits, and publication strategy. It also establishes the organization of the project.

The rationale behind the ACIA project is the increased recognition of the importance of regional assessments to complement the global picture provided by IPCC assessments (IPCC 1990, 1996), and the demand from policy makers and others for more relevant information. The goal of the ACIA is therefore to 'evaluate and synthesize knowledge on climate variability, climate change, an increased UV radiation <u>and</u> their consequences, and provide useful and reliable information to the governments, organizations and peoples of the Arctic region in order to support policy making processes and the IPCC's further work on climate change issues.'[20]

The implementation plan specifies that the ACIA will address '(i) the state of scientific knowledge and understanding of climatic processes across the region; (ii) the consequences of these changes across social and economic sectors relevant to the region as well as effects on ecosystems, biodiversity, and human health; and (iii) the foundations upon which nations and peoples of the region can adapt, adjust, cope, and/or take constructive advantage of the opportunities afforded by the changes.'

The ACIA has produced three documents: the main ACIA Scientific Report (ACIA 2005), a synthesis document providing a short and accessible version of major findings (ACIA 2004), and a Policy Document that relates ACIA scientific findings to the policy needs of the Arctic Council and its member governments, and offers recommendations for follow-up measures.

An Assessment Steering Committee (ASC) was to be in charge of the work of the ACIA, to provide 'guidance and oversight' of the assessment process. In addition to representatives of the AMAP, CAFF, and IASC, the lead authors of the chapters of the Assessment were represented in the ASC. Dr Robert W Corell of the American Meteorological Society was elected chair of the ASC. The Committee was also responsible for communication with the IPCC and other organizations. In 2000, a Memorandum of Understanding was signed between the ACIA and the IPCC, setting out the details of cooperation. A secretariat, funded by the USA, was established at the University of Alaska in Fairbanks. The secretariat was to play an active role in coordinating the work and providing various services to those preparing the assessments.

The ACIA was to operate on a 'lead author' strategy, whereby individuals with 'proven excellence in research, breadth of understanding of the subject, and demonstrated independence from major political or organizational interests' were appointed by the ASC to lead the work on specific chapters. To assist the lead authors the ASC appointed several contributing authors for each chapter – often as many as 10 to 20. Lead authors as well as contributing authors were in most cases paid by their countries. In general, the assessments of the status of various Arctic regions were carried out by scientists from the region in question. The lead author strategy and team-based approach of most chapters ensured a consistent treatment of the various regions.

In addition to the Implementation Plan, the Assessment Guidelines[21] provided technical guidance on timetables, usage of terms, scenarios, etc. The ACIA was able to deliver the popularized version of its report (ACIA 2004) on time – to the November 2004 ministerial in Reykjavik. The scientific report (ACIA 2005) was published almost a year late, due to technical difficulties in production.

An overriding concern in the design of the assessment was that the ACIA should be conducted by scientific experts in partnership with affected stakeholders. The assessment was to draw on existing research, rather than initiate new projects – the very idea of an assessment is to take stock of existing science as it stands in appropriately published form.[22] The contribution of stakeholders to the process was ensured by participation at various meetings.

The geographical scope of the assessment was to conform to a rather generous understanding of the geographical extent of the Arctic, including for instance northern Norway, Finland, and Sweden, as well as the Faeroe Islands. One reason for this was a need to include the sub-Arctic regions that are integral to the functioning of the Arctic climate system (ACIA 2004: 4). Another was the desire on the part of certain actors to be part of the work under the Arctic Council, to be seen and heard as Arctic nations. The ACIA area was divided into four sub-regions for the purpose of the assessment: the North Atlantic, Siberia/North Russia, the Bering Sea, and northern North America.[23]

Thematically, the scope of the ACIA was to cover the effects of emissions of greenhouse gases, as well as the effects of the depletion of stratospheric ozone. Ozone depletion is related to climate change in that a warming of the atmosphere will affect the recovery of stratospheric ozone negatively (ACIA 2004: 3). The assessment did not devolve into issues of mitigation and adaptation, which are now central to the work of the IPCC. The regions vary considerably with regard to the existence and quality of data. This created numerous challenges for the writing teams.

The baseline of the assessment was the so-called B2 scenario of climate change over the next century. B2 is a moderate scenario, and its predictions on global warming are below the mid-range among the various scenarios used by the IPCC. From this, the authors of the ACIA were to analyse the likelihood of change for 2020, 2050 and 2080 time-slices. The possibility of climate-induced change was to be indicated by the use of carefully defined wording (the 'lexicon'), where the probability of an occurrence was to be described on a scale ranging from 'very unlikely' to 'very likely', with 'unlikely', 'likely' and 'possible' in-between.

At an early stage, the ASC fleshed out an outline of the assessment report. Several chapters accounting for climate changes in the Arctic were to be followed up by chapters addressing the possible effects of climate change on various sectors: fisheries, forestry and agriculture, health and infrastructure. The final report contains 18 chapters, including the introduction and a final synthesis chapter.

Execution of the assessment was to take four years, from 2000 to 2004. Major meetings and workshops were arranged by the ASC during the assessment to bring the chapter writing teams together to produce the drafts. Later in the project, meetings focused on bringing lead authors together with a view to developing consistency and a sound division of labour between chapters.

An important aspect of the ACIA process was the attention paid to the authority and legitimacy of the scientific results. Transparency was seen as essential in this regard. All meetings were open, in the sense that those who asked for permission to attend would be allowed to do so. Also, lead authors were required to log all steps in the work in order to document the process. A thorough review process was conducted. The number of reviewers per chapter was higher than usual for scientific publications. In addition to an external review, various internal reviews were conducted.[24]

Communication and interaction with the IPCC were ensured by exchange of personnel. Scientists participating in the IPPCC took part in the work under several chapters of the ACIA scientific report (ACIA 2005). Now several ACIA authors are involved in the production of the next IPCC assessment report.

The broad outline of the ACIA work process, then, resembles that of the IPCC. The use of a lead author strategy, the distinction between science reports and policy documents, comprehensive peer reviews, emphasis on transparency, and firm governance of the process of producing the scientific report are important aspects. In particular, the ACIA demonstrated the importance of vigorous leadership, in the person of US meteorologist Robert W Corell. Corell came to play an important role not only in instigating and organizing the ACIA,[25] but also in communicating its findings to policymakers and the general public.

Major ACIA findings

The main message from the ACIA is that, taken as a whole, the Arctic region is experiencing rapid increases in temperatures, and that the temperature increases will have important consequences for Arctic ecosystems and societies. Moreover, climate changes are being experienced particularly intensely in the Arctic (ACIA 2004:8). The implications of

this are global, as climatic processes in the Arctic region have sig￼
cant effects on global climate.

The major findings as set out in the 2004 popular report (pp. 10–11) are as follows:

- The Arctic climate is warming at almost twice the rate of the rest of the world. An additional warming of 4–7°C over the next century will bring increased precipitation, shorter and warmer winters, and decreases in snow and ice cover.
- The warming of the Arctic has global ramifications. The reduction in snow and ice cover will increase the absorption of heat from the sun, leading to further increases in global temperatures. Glacial melt will contribute to raising the sea level considerably in the course of this century. Melting glaciers will also increase the freshwater supply to oceans, possibly affecting the thermohaline circulation that brings heat from lower latitudes to the polar regions.[26]
- Vegetation zones are likely to shift. In some parts of the Arctic, the treeline will move hundreds of kilometres northwards. Agriculture in some areas will also expand northwards, due to a longer growing season. Also forestry is likely to be enhanced.
- The diversity ranges and distribution of animals will be strongly affected by warming temperatures. The reduction in sea ice[27] will affect the habitat of polar bears and some seal species, pushing them towards extinction. The migration ranges of several marine as well as terrestrial species will move northwards, bringing new species into the Arctic. Marine fisheries in the region are likely to become more productive, as temperature increases will bring higher biological production.
- Increased frequency and intensity of storms pose challenges to coastal communities in particular. Coastal erosion is likely to increase, as is the risk of flooding of low-lying areas.
- Reduced sea ice is very likely to enhance transportation and access to some natural resources. The navigation season is very likely to be longer, and summertime trans-Arctic shipping may be feasible some decades from now. Reduced sea ice is likely to enhance the conditions for offshore petroleum activities. Conflicts over sovereignty, security, and rights to resources are likely to increase.
- Reductions in permafrost[28] and thawing of ground will affect travel on ice and tundra. As a consequence, buildings, roads, and other infrastructure are likely to be destabilized, requiring substantial investment for their maintenance.

people and their communities are likely to be exposed
)f these changes, depending on where in the region
ose depending on a subsistence economy will be most

ise gases in the stratosphere inhibit the improvement
ᴗₗ ᴛɦe ozone layer. As a consequence, UV and Uvb radiation can
be expected to remain high for several decades.[30] UV radiation can
also have disruptive effects on photosynthesis and on young fish.

Climate change occurs in the context of various other changes, such
as globalization, increases in pollution, and land-use change. Such
changes may combine to amplify the impacts of climate change. The
fact that climate change in the Arctic occurs in the context of other
major changes in the region, posed a methodological challenge for the
chapters on the impacts of climate change on economic sectors in particular, as the question of how to identify and assess the variation that
can be related to a climate signal is very complex indeed.[31]

An important aspect of these findings is that they do not in all cases
apply universally to the Arctic as a whole. What constitutes the most
important economic effects appears to vary from region to region, as
presented in the ACIA popular report (ACIA 2004: 18–19):

In the North Atlantic sector (East Greenland to Northwest Russia),
for instance, the positive economic effects for fisheries and petroleum
activities are emphasized. For the sector centring on the Bering Sea
(Chukotka to Western Canadian Arctic), the disruptive effects of thawing permafrost on infrastructure are highlighted. In the Russian sector
(Siberia) the economic opportunities that retreating sea ice provides
for shipping and offshore activities are in focus. For the sector ranging
from the Central Canadian Arctic to West Greenland, both positive
and negative economic effects are identified. For instance, reduced sea
ice provides opportunities for shipping, on the one hand, but also increased risks of oil spills.

Other outcomes

The ACIA identified several issues that need further scientific attention.
The 2004 report, while pointing out that it represents the beginning
of a process (p. 122), also contains several science recommendations.
While Arctic wide impacts were extensively addressed, the report recognizes that when it comes to economic impacts and sub-regional impacts, 'greater development of such estimates must be a future priority
task' (p. 122).

Three priority tasks for future research are suggested that add societal impacts of climate change: sub-regional impacts, socio-economic impacts, and the assessment of vulnerabilities. This research agenda has to be accompanied by a range of activities relating to long-term monitoring, process studies, climate modelling and social impact assessments.

Impacts: effective mobilization at the regional level?

The issue of regime effectiveness has spurred a vast academic literature (Young 1999). Important distinctions exist between various ways of conceiving of effectiveness (Underdal 1992). A demanding definition sees effectiveness as the resolution of a given problem, the problem that motivated the establishment of the regime in question. For the case at hand here, such a definition of effectiveness is not very relevant. Current efforts to confront global warming under the Kyoto Protocol are timid,[32] and global emission levels will continue to increase over the next decades. Even if emissions were to be cut to 1990 levels today, the effects on the climate system would take centuries to emerge.

Another understanding of the concept of effectiveness is that a given regime can be termed effective when it induces change in actor behaviour in directions that can be assumed to promote resolution of the problem. This is a relevant definition here, and would indicate that we should look for actions by the Arctic 8 nations that in some way or other affect the climate problem.

The 'climate problem' is a complex one, and we should distinguish actions geared towards learning more about the issue from actions aimed at reducing emissions. Actions aimed at adaptation to the problem would constitute a third category. The first type of actions would consist in calls from political bodies for more information and new knowledge, while the second would mean decisions on for example emissions reductions and timetables.

Effectiveness

Effectiveness in the case at hand, then, is about the initial and modest, preliminary steps taken to understand and address an environmental problem. A precondition for action to remedy a problem is the recognition and understanding of it (Keohane *et al.* 1993; Andresen *et al.* 2000). Here the ACIA plays an important role, by raising awareness and producing knowledge that can enhance our understanding of the climate problem.

te effect of the ACIA was in the 2004 ministerial dec-
Arctic Council and in the Policy Document emanat-
rocess. In the months leading up to the 2004 Arctic
:rial Meeting there had been much speculation whether
ne ACIA policy recommendations would be achieved.
tention was that a group of experts should develop the
Policy Document. The controversy generated by the US opposition to
the idea brought the issue to the table of the Senior Arctic Officials,
who negotiated a text that is referred to as the Policy Document. US
opposition to the idea of a policy document was adamant, and the
process to produce this broke down for a period. However, in the end,
consensus was reached and the Senior Arctic Officials were able to
provide the Arctic Council Ministers with a Policy Document, based on
the elements in the Ministerial Declaration and SAO Report to Ministers. The recommendations in the Policy Document call for member
states to concentrate their domestic efforts on mitigation, adaptation,
research, observations, monitoring and modelling, as well as outreach
activities. At the same time, it directs the Arctic Council Working
Groups to consider the findings of the ACIA in their activities and in
developing follow-up actions

As to awareness, the ACIA drew the attention of decision makers
almost as soon as the idea of an Arctic climate impact assessment was
launched. The 2000 Arctic Ministerial provided the mandate for the
project to be undertaken in the first place. The ACIA attracted considerable attention in the media as well as in various environmental forums.
It was presented at the 2002 World Summit on Sustainable Development (WSSD) and received significant attention from actors outside
the Arctic region. At the 2004 Arctic Council ministerial meeting the
ACIA was the major issue, bringing global media to an Arctic Council
meeting for the first time. Important international news outlets like the
BBC and the *New York Times* have had high-profiled articles about the
ACIA, and its findings have been widely broadcast.[33] An examination
of the media impact of the ACIA compared to the IPCC in the USA
(Tjernshaugen and Bang 2005), finds that the ACIA report attracted
more attention than the IPCC. A major reason was the effort invested
in communicating the results of the assessment.[34] Also, the ACIA findings emerged as less controversial than those of the IPCC (2001), and
a sharper distinction was drawn between the production of the science
report and the policy document than in the IPCC.

Not least in the USA has this been important, where the scepticism of the public as well as the government to the findings of the
IPCC is deeply rooted. The ACIA put a face on the climate issue in
the USA, by providing numerous examples of climate-related events in

hm?.

Alaska as well as indicating likely future hazards. Also, Washington's inelegant diplomatic manoeuvres in the production of an ACIA Policy Document, where US diplomats for some time practically sabotaged progress, spurred scepticism in US public opinion as to the real motives of the administration (Prestrud 2004).

The ACIA has also contributed to enhance the understanding of US politicians of the climate issue. Several prominent senators participated in an excursion to Svalbard, an archipelago located at about 80° N. latitude, to the north of the Norwegian mainland. Here, they had several days of lectures and demonstrations of climate impacts on the local environment, and went home convinced of the seriousness of the issue. The fact that a prominent Republican senator, John McCain, participated and spoke convincingly on the issue, indicates that some impact was produced.[35] Also, a hearing on climate change was held in the US Senate as a consequence of the ACIA findings (Tjernshaugen and Bang 2005:10).[36]

The Ministerial Declaration from the fourth meeting of the Arctic Council (2004) contains a section on climate change in the Arctic.[37] The Council notes the ACIA findings and acknowledges that such findings and the underlying scientific assessment will help inform governments as they consider and implement policies relating to climate change. The Declaration also endorses the ACIA policy recommendations for research, mitigation, adaptation, monitoring, and outreach as stated in the SAO Report to Ministers. The latter probably contains the most progressive language on the climate issue used by the current US administration, in terms of recognizing the gravity of the problem as well as steps to be taken. Furthermore, the Declaration encourages member states to take effective measures to adapt to and manage impacts of climate change through enhancing Arctic residents' access to information, to decision makers and to institutional capacity building.

An important outcome in terms of effectiveness, therefore, is the engagement of the ministers from the Arctic 8 countries in the issue. The nature and content of the Declaration goes surprisingly far in recognizing the problem and suggesting courses of action, given the US reservations on the issue. It should also be recalled that the USA was in fact the lead country on the project, hosting the secretariat and offering vigorous leadership. This might be interpreted as an attempt to control the process to make the outcome conform to US interests. However, given the way the ACIA process unfolded and its outcomes, a more plausible explanation is that the US involvement was dominated by actors who were profoundly committed to the scientific quality of the project and to unfettered communication of its results.[38]

The impact on policy makers' perceptions and understanding of the climate issue is perhaps the most significant ACIA outcome thus far. In the USA in particular, the climate issue is now widely recognized and firmly placed on the political agenda. The ACIA has contributed substantially to this, and this will in turn have impacts on the global climate regime.

As to direct effects on the global climate regime, impacts can be traced in two realms: on its knowledge production wing, the IPCC, and on the arena constituted by the UNFCCC Kyoto protocol.

The 2004 Arctic Council Ministerial Declaration encourages relevant research bodies to take into account the scientific recommendations of the ACIA in the planning and implementation of their own programmes. Since the IPCC provides the knowledge base on which the work under the Kyoto Protocol is built, an essential contribution to the climate regime lies in the improvement in scientific knowledge that the ACIA represents. The basis for this work has been assessments carried out by the IPCC. In turn, future IPCC assessments will necessarily draw on the work of the ACIA. The next IPPCC assessment report, due in 2007, will have a separate chapter on impacts of climate change in the Arctic. The ACIA has also fed into other international scientific bodies like the World Climate Research Programme, as well as numerous national programmes. In this way, it has served to boost the work of the IPCC, and in turn, the political wing of the global climate regime.

As to the relationship to the global climate regime, the 2004 Arctic Council Ministerial Declaration speaks of the 'need to consider the findings of the ACIA and other relevant studies in implementing their commitments under the UNFCCC and other agreements', as well as the need to promote global, national and local awareness of the ACIA and follow-up activities. The follow-up has involved some coordinating activities among the Nordic countries in the preparation of and during the global talks. During the 2005 ministerial of the Kyoto Protocol, a side event in the form of an 'Arctic Day' was arranged where the ACIA and its findings were presented. Several ministers also emphasized the challenges of the Arctic region in their interventions.[39]

Mobilization

The concept of mobilization refers to how actors and groups become involved (or not) in an issue area, and whether and how changes in the set of actors have an impact on developments in the issue area in question. The climate issue has for long ranked high on the agendas of the academic world and in the environmental community. Its salience to society at large is more recent, and in the Arctic context it has attracted broad attention in the last few years.[40]

In science, the ACIA contributed to mobilization and formation of new networks among scientists, from a range of countries and academic disciplines. This served to enhance both international collaboration in climate research (indispensable in that regard), and to promote cooperation across disciplinary boundaries. In complex issue areas like climate change, multi-disciplinarity is a precondition for advances in science, in particular for studying the impacts of change. This is probably also important in the relationship to the IPCC, as the Arctic climate issues now have their own scientific constituency with shared perceptions and ideas about the region.[41]

International cooperation in Arctic science has a long tradition, in recent years promoted by the International Arctic Science Committee (IASC), established in 1988. IASC, formally an NGO operating under the auspices of the International Council for Scientific Unions (ICSU), has initiated a large number of international research projects and programmes. But no IASC initiative has commanded such a wide following and attracted as much attention. In the end more than 300 scientists from 15 countries participated in the ACIA.

Also other actor groups were mobilized or saw on-going mobilization enhanced by the climate issue – indigenous people in particular (see Chapter 3 by Wilson and Øverland), who saw their concerns addressed in a comprehensive manner. Three chapters relating to indigenous issues are found in the ACIA scientific report (Huntington and Fox 2005; McCarthy and Long Martello 2005; Nuttall 2005). The latter proved somewhat controversial among indigenous people, and one case study was withdrawn from the chapter. Environmental NGOs remained surprisingly oblivious to the project, and only the WWF had any notable presence over time in the process.

Also economic groups like the fishing and shipping industries were involved in the ACIA process. While marginal to the ACIA itself, their involvement probably represented the first opportunity for them to hear about circumpolar cooperation and get an early, systematic introduction to various aspects of the impacts of climate change in the North. An important mobilization outcome of the ACIA, therefore, has been the enhanced awareness of circumpolar cooperation. What had been, at best, a jumble of acronyms became known also to those outside government circles. In this sense, the ACIA has brought fame to the Arctic Council.

Region building

An important aspect of circumpolar cooperation as it has developed over more than a decade is the search for a shared Arctic identity. Given the vastness and diversity of the region, the common denominators are, however, few and far between. With the advent of the climate issue on the scientific and then political agendas, a common issue emerged – one in which actors in all Arctic nations had a shared perception and interest in pursuing. The significance of the ACIA in this respect is to draw attention to the issue and provide a focal point and arena for mobilization of knowledge around the issue.

To the Arctic Council, the climate issue is important also because it gives the body a role on an issue perceived as important to all participating countries. As pointed out above, the climate issue and the ACIA have also served to make the Arctic Council and its work programs better known among people in the Arctic region. This has served to enhance the legitimacy of the Arctic Council among policy makers, and facilitated an expansion of its agenda. This is shown by the decision to recommend an ACIA follow-up, as well as the on-going study on petroleum activities in the circumpolar region under the AMAP.

The *modus operandi* of the Arctic Council, where the country (or countries) that finances a project and acts as lead country can dictate its agenda, may, however, undermine this aspect of region building, if a country chooses to utilize a programme for its own ends, contrary to the interests of other countries. In the case of the ACIA, this was in fact an issue, given the position of the current US administration on climate change. The behaviour of US officials in negotiations over the Policy Document substantiates this. However, most countries paid for the participation of their own scientists, and the US scientific leadership of the process was not swayed by the political concerns of its government, but took an independent stance informed by the science produced under the project.

A further aspect of region building is that it is a process that feeds on itself, in the sense that increased activity and recognition at that level makes it more attractive for others to join in the process and take part in the activities. Especially for small nations that have difficulty in gaining visibility in major international forums, the Arctic Council with its programmes represents an arena for displaying nationhood and statesmanship. This pertains particularly to countries that do not have indigenous populations that are highly visible in the Arctic Council itself. The ACIA served as a vehicle for such ambitions for Iceland (240,000 inhabitants) and the Faeroes (40,000 inhabitants) in particular.

In addition to the substantive outcomes of the ACIA project, an important aspect is that an enormous amount of data on Arctic ecosystems and communities was collected. Huge efforts were invested in processing data to make them compatible across geographical regions and scales. The ACIA has therefore contributed to an improved understanding of the Arctic region as a whole, its dynamics and development, and has yielded a comprehensive Arctic data inventory in the process. The ACIA fisheries chapter (Vilhjamsson and Hoel 2005) provides the first real account of Arctic fisheries for the region as a whole, and the same can be said for other issues as well. The development of these Arctic data inventories has acted to define a regional approach likely to shape the scientific approach to the region for a long time to come.

By the same token, the ACIA has also contributed to strengthening the regionalization of Arctic science. It led to the formation of several science networks centring on the various chapters of the Scientific Report (ACIA 2005), which in several instances have continued to exist post-ACIA, among other things to produce special issues of scientific journals and to develop proposals for new ventures in Arctic science. The ACIA has thus served to reinforce the position of the International Arctic Science Committee, and instigated networks to be mobilized in other contexts, like the Arctic Human Development Report (AHDR 2004) and the upcoming International Polar Year.

Conclusions

The Arctic 8 countries are responsible for almost half the global emissions of climate gases. At the same time, the Arctic regions of some of these countries are among the regions in the world likely to be most affected by climate change. Collectively, they therefore have a common interest in confronting the climate challenge. That is a task vested in the global climate regime, where reduction targets and timetables are laid down in the Kyoto Protocol. Among the Arctic countries we find the most ardent supporters of this regime, as well as its pre-eminent adversary, the United States.

As a group, the Arctic 8 countries have increased their emissions substantially since 1990. This is contrary to the general trend among developed countries, which have seen a decrease of 5.9 per cent in the period 1990–2003.[42] In the global climate regime, therefore, the performance of the Arctic 8 countries leaves much to be desired. Great leaps forward in performance are not to be expected in the short term, as the USA is unlikely to become a party to the global regime in the foreseeable future. In this regard it is perhaps regional cooperation within the European Union that holds most promise.

In the context of Arctic circumpolar cooperation, work on the issue in relation to climate change has also served to enhance our knowledge of the climate problem and its impacts on ecosystems and societies. This is an institutional response that does not match the severity of the problem, but this aspect is not special to the Arctic. The Arctic Climate Impact Assessment, performed over about five years, brought an unprecedented level and scale of scientific cooperation to the Arctic. Through an elaborate process involving more than 300 scientists from 15 countries, a 1,042-page tome of a scientific report (ACIA 2005) was produced, presenting the state of knowledge on a broad range of issues relating to climate change in the Arctic. In addition, an overview report cast in an easily accessible language and with innovative graphic presentations was produced, as was a policy document.

ACIA effectiveness is a matter of impact of knowledge on decision makers. The pathways of influence in such settings are usually complex (Andresen *et al.* 2000). The main message here is that the ACIA contributed substantively to greater understanding of the climate problem among decision makers. This is particularly important in the USA, given the country's position on the issue and its non-membership of the global climate regime, together with the fact that it is responsible for a quarter of the global output of climate gases.

Various factors contribute to explaining the effectiveness of the ACIA. One is the quality of the science involved, which produced a very robust and comprehensive assessment of the status of scientific knowledge on the climate issue in the Arctic. A further factor is the consensus in the scientific community on the main message of the project – that the Arctic is warming rapidly, and that this has numerous consequences on enormous scales of magnitude. Still another explanation is that controversy over the content of the policy document was kept out of the scientific process, due to the organization of the ACIA project, which maintained a clear distinction between scientific work and policy making recommendations. Still another factor in ACIA effectiveness is its success in communicating its findings to a broad audience.

The ACIA contribution to the global climate regime is more elusive. The question of changes in state behaviour to adopt measures to curb emissions of climate gases more aggressively is probably best thought in terms of the post-Kyoto (2012 onwards) period. Its current impact on US decision makers in particular is important in this regard. And some action can already be observed: some of the Arctic states are coordinating their work under the Kyoto Protocol. One effect that can readily be observed concerns the work of the scientific wing of the global climate regime, the IPCC. The 2007 Assessment Report will have a chapter on impacts of Arctic climate change. Emerging follow-up activities are also significant, as they will contribute to greater knowledge on the issue.

In terms of mobilization, the climate issue has demonstrated that great-power backing is not a necessary condition for a programme to have an impact. The ACIA has served as a mobilizing force in the Arctic despite the scepticism of the current US administration. The mobilization of a large network of scientists, including people and institutions from non-Arctic countries, is particularly important. The International Arctic Science Committee (IASC) was a key actor in this regard. The ACIA has resulted in a greater interface of cooperation between the IASC and science on the one hand, and the Arctic Council and politics on the other.

The climate issue as such has emerged as a strong vehicle for regional identity, in particular among indigenous people. The reasons are complex, but being jointly against something certainly contributes to the creation of a shared identity. Among other groups, the ACIA is significant perhaps first and foremost by having contributed to making the Arctic Council and its work better known among the people of the region.

The ACIA has therefore served to enhance the legitimacy of the Arctic Council in the region. As to regional cooperation in science, the overview report identifies three priority tasks that address societal impacts of climate change in particular: sub-regional impacts, socio-economic impacts, and the assessment of vulnerabilities. This research agenda will have to be accompanied by a range of activities relating to long-term monitoring, process studies, climate modelling and social impact assessments.

Notes

1 The Convention text can be found at: http://unfccc.int/essential_background/convention/background/items/2853.php
2 The Protocol and related texts can be found at: http://unfccc.int/essential_background/kyoto_protocol/background/items/1351.php
3 Canada, Denmark, Finland, Iceland, Norway, the Russian Federation, Sweden and the USA
4 The outgoing radiation from the Earth has a wavelength that is more easily trapped by the 'layer' made up of these gases, than the incoming radiation from the Sun.
5 'Joint implementation' works by an Annex I country assisting other devel-

ıntries to reduce emissions. The 'clean development' mechanism
the same mechanism applied to developing countries.
:o force of the treaty required ratification by 55 per cent of the
es as well as the ratification of Annex I states responsible for 55
of the greenhouse gases emissions.
7 Under the Clinton Administration, the USA signed but never ratified the
   Kyoto Protocol.
8 See http://ipcc.ch for further information on the work of the IPCC.
9 Figures from Key GHG Data, at http://unfccc.int/essential_background_
   publications_htmlp
10 A much-quoted report in *The Guardian* alleged that the US decision followed directly from pressure from Exxon-Mobil (8 June 2005).
11 Critical issue areas in this regard – like energy, environmental and transportation policies – are areas where the EU and its member states have
   shared competences.
12 The system is based on national allocations of tradable emissions allow-
   ances that will be reduced over time to meet Kyoto targets. A community
   burden-sharing scheme defines the objectives to be reached by each member country.
13 As quoted by Vogel and Bretherton (2006:15).
14 The European Union was instrumental in securing Russia's ratification.
   EU support for Russian membership of the World Trade Organization,
   for which Russia had been negotiating for several years, was made contingent upon the country's accession to the Kyoto Protocol. See http://www.
   peopleandplanet.net/doc.php?id=2237.
15 An early instance of circumpolar cooperation in environmental affairs is
   the 1972 treaty on protection of polar bears (Fikkan *et al.* 1993).
16 See http://www.iasc.se.
17 World Climate Research Program.
18 Report of a meeting and workshop to plan a study of the impacts of climate change on Arctic regions, 28 February–1 March 2000, Washington,
   DC
19 Barrow Declaration, para. 3, available at: http://www.arctic-council.org/
   files/infopage/75/bar_decl.pdf
20 ACIA Implementation Plan, p. 9, available at: http://www.acia.uaf.edu/
   PDFs/Implementation-Plan.pdf
21 Available at: http://www.acia.uaf.edu
22 It soon turned out that additional research was required, and ACIA spurred
   many new projects which contributed substantially to the outcome.
23 These regions vary considerably with regard to the existence and quality
   of data. This created numerous challenges for the writing teams.
24 A special feature of the review process was that lead authors were required
   to document exactly how they had responded to reviewers' comments.
25 Corell was a member of the IASC Executive Committee and was instrumental in developing the idea of an assessment of impacts of climate
   change in the Arctic.
26 The Greenland icecap experienced a 16 per cent increase in melting in the
   1979–2002 period (ACIA 2004:13).
27 The extent of sea ice has decreased 15–20 per cent over the last three dec-

ades. This trend is expected to accelerate, with 'a near total loss of summer ice in summer projected for late in this century' (ACIA 2004:13).

28 Permafrost is warming up, and the layer that thaws is increasing. The southern limit of permafrost is projected to travel several hundred kilometres northwards.

29 People who depend upon animals whose habitat is on sea ice are likely to be most severely affected. This may lead to a change to Western-style diets, with increased risk of the associated diseases (diabetes, obesity, cardiovascular ailments).

30 Young people in the region will receive a dose of UV radiation 30 per cent higher than that of any previous generation.

31 ACIA paid some attention to this, and the issue is addressed in for example the fisheries chapter of the science report (ACIA 2005).

32 To make substantial progress in reducing the negative consequences of global warming, the increase in global mean temperatures has to be kept below 2°C. Achieving this requires that emissions levels be brought down 60–80 per cent by 2050, relative to 1990 levels. In that perspective, fulfilling the Kyoto Protocol is merely a beginning, as it will bring only a 5.2 per cent reduction by 2012.

33 Including presentations in popular shows like '60 minutes' and 'the Late Show' broadcast nationwide in the USA, and *Rolling Stone* magazine.

34 A professional author wrote the overview report (ACIA 2004): considerable time and effort was devoted to graphical material, and presentations focused on a few, specific examples that were easy for the non-expert to understand and relate to.

35 McCain, a Republican representative, has a history of promoting a stronger domestic policy on curbing emissions of climate gases. Other participating senators included Hillary Clinton.

36 On 16 November 2004. Previously, three hearings had been held on the third IPCC assessment report.

37 Reykjavik Declaration, available at: http://www.arctic-council.org/

38 As the success of ACIA became evident, the leader of the project, Dr Robert W Corell, was increasingly attacked in US media and in internet-based communications.

39 See, for example, the speech by Norway's ministry of the environment to the 2005 Ministerial of the Kyoto Protocol, at http://odin.dep.no/md/norsk/aktuelt/taler/minister/

40 While ACIA is an important driver in that development, work under the IPCC that predated, and contributed to the initiation of the ACIA, has also been important in that regard. (See Anisimov and Fitzharris 2001)

41 The idea of a continuation of ACIA, focusing more on impacts and socio-economic aspects, is under discussion. Norway has established a domestic ACIA II project.

42 Press release, UNFCCC, Bonn, 17 November 2004.

References

ACIA (2000) Implementation Plan. Available at: http://www.acia.uaf.edu/PDFs/Implementation-Plan.pdf

4a) *Impacts of a Warming Arctic* (popular version of scientific re-ambridge: Cambridge University Press.

4b) Policy Document. Available at: http://www.acia.uaf.edu/PDFs/olicy_Document.pdf

ACIA (∠υυ5) *Arctic Climate Impact Assessment*, Cambridge: Cambridge University Press.

AMAP (1997) *Arctic Pollution Issues – A State of the Arctic Environment Report*, Oslo: Arctic Monitoring and Assessment Programme (AMAP).

AMAP (1998) *AMAP Assessment Report: Arctic Pollution Issues*, Oslo: AMAP.

AMAP (2004) AMAP *Assessment 2002 – Radioactivity in the Arctic*, Oslo: AMAP.

Andresen, S., Skodvin, T., Underdal, A. and Wettestad, J. (2000) *Science and Politics in International Environmental Regimes*, Manchester: Manchester University Press.

Anisimov, O. and Fitzharris, B. (2001) 'Polar regions' in J.J. McCarthy, O.F. Canziani, N.A. Leary, D.J. Dokken and K.S. White (eds.), *Climate Change 2001: Impacts, Adaptation, and Vulnerability*, Cambridge: Cambridge University Press, pp. 801–842.

Arctic Human Development Report (AHDR) (2004), Akureyri: Stefansson Arctic Institute.

Benedick, R.E. (1991) *Ozone Diplomacy: New Directions in Safeguarding the Planet*, Cambridge, MA: Harvard University Press.

Bolin, B. (1997) 'Scientific assessment of climate change' in G. Fermann (ed.), *International Politics of Climate Change*, Oslo: Scandinavian University Press, pp. 83–109.

Council of Ministers of the European Union (2005) *Environment 2647 Session 6693/05*, Brussels, 10 March.

Dessler, A.E. and Parson, E. (2006) *The Science and Politics of Global Climate Change*, Cambridge: Cambridge University Press.

Fikkan, A., Osherenko, G. and Arikainen, A. (1993) 'Polar bears: the importance of simplicity', in O. Young and G. Osherenko (eds.), *Polar Politics*. Ithaca, NY: Cornell University Press.

IPCC (1990) *Climate Change: The IPCC Scientific Assessment*, edited by J.T. Houghton, G.J. Jenkins and J.J. Ephraums, Cambridge: Cambridge University Press.

IPCC (1996) *Climate Change 1995: Second Assessment Report on Climate Change*, Cambridge: Cambridge University Press.

IPCC (2001) *Climate Change 2001: Impacts, Adaptation, and Vulnerability*. Cambridge: Cambridge University Press.

Huntington, H. and Fox, S. (2005) 'The changing Arctic: indigenous perspectives' in *ACIA 2005*, Cambridge: Cambridge University Press, pp. 61–98.

Keohane, R.O., Haas, P.M. and Levy, M.A. (1993) 'The effectiveness of international environmental institutions' in P.M. Haas, R.O. Keohane and M.A. Levy (eds.), *Institutions for the Earth: Sources of Effective International Environmental Protection*. Cambridge, MA: MIT Press, pp. 3–26.

Klein, D. (2005) 'Management and conservation of wildlife in a changing Arctic environment' in *ACIA 2005*, Cambridge: Cambridge University Press, pp. 597–548.

McCarty, J.J. and Long Martello, M. (2005) 'Climate change in the context of multiple stressors and resilience' in *ACIA 2005*, Cambridge: Cambridge University Press, pp. 945–988.

Nuttall, M. (2005) 'Hunting, herding, fishing, and gathering: indigenous peoples and renewable resource use in the Arctic' in *ACIA 2005*, Cambridge: Cambridge University Press, pp. 649–690.

Osborn, T. and Briffa, K.R. (2006) 'The spatial extent of 20th-century warmth in the context of the past 1200 years', *Science*, 311 (5762): 841–844.

Prestrud, P. (2004) 'Arctic Climate Impact Assessment', *Cicerone*, 6: 3.

Tjernshaugen, A. and Bang, G. (2005) *ACIA og IPCC – en sammenlikning av mottakelsen i amerikansk offentlighet*, CICERO Report 2005:04, Oslo: Cicero.

Underdal, A. (1992) 'The concept of regime "effectiveness"', *Cooperation & Conflict*, 27 (3): 227–240.

Victor, D.G. (2004) *Climate Change: Debating America's Policy Options*, New York: Council on Foreign Relations.

Vilhjamsson, H. and Hoel, A.H. (2005) 'Fisheries and aquaculture' in *ACIA 2005*, Cambridge: Cambridge University Press, pp. 691–780.

Vogler, J. and Bretherton, C. (2006) 'The European Union as a protagonist to the United States on climate change', *International Studies Perspectives*, 7: 1–22.

Young, O. (1998) *Creating Regimes – Arctic Accords and International Governance*, Ithaca, NY: Cornell University Press.

Young, O.R. (ed.) (1999) *The Effectiveness of International Environmental Regimes: Causal Connections and Behavioral Mechanisms*, Cambridge: MIT Press.

7 Oil, gas and the environment

Kristine Offerdal

Introduction

Technological advances and a warmer climate have made rich hydrocarbon reserves in the Arctic increasingly accessible. The opportunities of social and economic developments in the region have been enhanced, but a future escalation of oil and gas extraction and transportation can also involve potential threats to the Arctic environment.

Even though the maritime environment is affected by onshore activities and the related transportation of oil and gas, the recent increase in attention to the environmental effects of hydrocarbon activities is due primarily to the renewed interest in the potential of Arctic offshore developments. Since these activities are still relatively modest, the ambition of this chapter will be to assess the *preparedness* of Arctic institutions, especially the Arctic Council, to meet the new challenges linked to offshore hydrocarbon activities. As yet, the Barents Euro-Arctic Region (BEAR) and the Council of the Baltic Sea States (CBSS) have not prepared policy documents on oil and gas issues within an Arctic environmental framework, so they will not be included in our discussions here.[1]

The following questions will be raised: How prepared is the Arctic Council to address the environmental challenges connected to oil and gas developments in the Arctic, and what effects may be identified from this work (*effectiveness*)? To what degree do the United States, the EU and the Russian Federation use the Council as a forum for addressing their needs and priorities within the issue area? Has the Arctic Council improved the ability of new groups to participate and make their voices heard on environmental issues related to hydrocarbon developments (*empowerment*)? Lastly, does this particular forum contribute to the identification of the Arctic as a distinct region on the

international political agenda through its work on I
(*region building*)?

Six interviews and several informal conversation
the data material, together with Internet searches
ing.[2] For practical reasons, most examples and da
Norway. This has obvious implications for genera
mentation level, although generalizations will be made
regarding the Council's future problem-mitigating potential. First,
however, I will give a presentation of Arctic hydrocarbon activities and
one main perception of related environmental challenges.

Arctic hydrocarbon activities and related environmental challenges

The long-expected Arctic 'energy rush' has not yet become a reality but
recent developments have made an escalation of hydrocarbon-related
activities more probable. These developments include factors such as a
warmer Arctic with less ice; the depletion of oil and gas in more south-
erly fields of Arctic oil- and gas-producing countries; continuing unsta-
ble political developments in producing regions elsewhere; the need
for greater security in energy supplies; high oil and gas prices; better
technology and renewed interest in the Arctic as an energy region on
the part of political as well as industrial actors.

Estimate from the United States Geological Survey (2000) indicate
that about 24 per cent of the world's remaining undiscovered hydrocarbon resources may be found in Arctic areas. Although this is a disputed figure, we may assume that the Arctic contains a substantial portion of the world's oil and gas reserves.[3] British Petroleum's Northstar,
located north of Prudhoe Bay at Seal Island, was the first offshore oil
project in the Arctic Ocean (Nowlan 2001: 49), with production start
in November 2001. It was also the first approved plan in Arctic Alaska
to use a sub-sea pipeline (Kitsos 2000). The first offshore gas field to
be developed in the European Arctic will be the Norwegian Statoil
project 'Snøhvit' in the southern parts of the Barents Sea, with its 5.7
trillion cubic feet (tcf) of natural gas reserves. Regular gas deliveries
are expected from December 2007. Production start for the Russian
oil field Prirazlomnoye, in the Pechora Sea, is scheduled for 2007. The
gigantic Shtokman gas field in the Barents Sea holds 113 tcf of proven
reserves,[4] or about twice the known gas reserves of Canada. It has an
official estimated production start in 2010, although expectations are
that this lies further into the future (OECD/IEA 2004: 311).

the next few decades, gas production from relatively undeveloped [fie]lds and new areas is expected to offset fully the declines in the main [es]tablished basins in the USA and Canada (OECD/IEA 2004: 150). The undeveloped Mackenzie River Delta in northern Canada and the Alaskan North Slope are included here.[5] Major Russian gas-producing fields in West Siberia are also in decline, but the Zapolyarnoye field is expected to compensate for much of this (OECD/IEA 2004: 311).[6]

Despite these ongoing and planned activities, the likelihood of significantly greater production in the very near future can be questioned. The issue of oil and gas in the Arctic has been on the political agenda of Arctic states before.[7] One major obstacle has been the lack of infrastructure in the sparsely populated Arctic region. Moreover, with regard to Russia, high world oil prices and the current tax regime and macroeconomic conditions seem to favour short-term projects and increased production from fields already in operation (Kim 2005: 370). On the other hand, new fields will eventually be developed, so it is reasonable to inquire into the related environmental challenges.

One activity that has already increased substantially is oil tanker traffic from Northwest Russia. Monthly cargo volumes from Russia through the Norwegian Economic Zone, for instance, have almost tripled – from 4.3 million tons in 2002 to 11.8 million tons in 2004 (Årøy 2005: 47).[8] Routes through the Arctic dramatically shorten the distances between commercial regions and trade centres: the Northern Sea Route, for example, is up to 40 per cent shorter through the Arctic than through the Suez Canal (ACIA 2004: 83). Reduced sea ice is expected to prolong the navigation season from the current 20–30 days to 90–100 days per year in 2080. However, considerable icebreaking capacity is needed to operate the NSR. Ragner (2000: 577) notes that the lack of such capacity represents a major future bottleneck for NSR operations.[9]

The primary sources of petroleum hydrocarbon inputs from marine shipping on a worldwide basis are discharges from fuel-oil sludges and machinery-space bilges and from oil tanker operations (IMO 1993, AMAP 1998). Although tanker accidents contribute only a small percentage of total inputs of oil to the sea worldwide, they remain the focus of much public attention. At present, environmental impacts from operational discharges of oil and other wastes in the Arctic are low. Acute pollution due to accidental events has not often been reported either (PAME 2000: 4). Nevertheless, in parts of the Arctic, sea ice, moving icebergs and even small iceberg fragments constitute considerable risks to super-tankers transporting oil or LNG (Jumppanen 1990:

79). Finally, expected continued increases in oil shipping in the Arctic will increase both the amount of oil from chronic small discharges and the probability of major tanker accidents.[10]

In addition to the increased risk of accidents, primary concerns associated with major new developments include the difficulties of taking remedial measures in such harsh environments. Other major concerns relate to the potential effects of spilled oil on commercial fisheries and the loss or alteration of habitats (AMAP 1998: 662). Moreover, the effects of oil pollution may be more severe and persistent in the Arctic environment than elsewhere (AMAP 1998: 661).[11] A common concern is also that the Arctic environment is particularly sensitive to pollution. Much of its human population and culture is directly dependent on the health of the region's ecosystems.[12] Consequently, a main perception is that oil and gas extraction and transport require particularly strict environmental standards in the Arctic.

Here it should be noted that various actors' perceptions of our ability to deal with environmental challenges may vary.[13] Some actors tend to emphasize the possibilities that the new developments provide and speak of the challenges as 'challenges to be overcome'. They have a strong faith in the industry's ability to develop technical solutions to the challenges. In this chapter, however, the analysis will be based on the fact that one main perception is that increased oil and gas activities in the Arctic will lead to environmental challenges which must be dealt with. In the following sections we will see how the Arctic Council perceives of and deals with some of these challenges. The identification of the perceptions of threats to the Arctic environment from hydrocarbon activities must be kept in mind throughout the discussion. First, however, let us look at the nature of the Council, and define what would be considered an effective Arctic Council, taking into consideration the special character of the issue area of oil and gas.

What can we reasonably expect?

The Arctic Council is an organization with a relatively low degree of formalization. While there is not a permanent secretariat, there are formalized rules as to how often the member states are to meet; it is clear who are the member states and who are the observers, as well as their mandates at meetings; and there are clear regulations on decision making within the Council (by consensus). On the other hand, once decisions have been made, there are few regulations on how they are to be followed up. Member states engage in and contribute financially to

Arctic Council projects according to their own interests and resources. These features give important processes an ad hoc character and underline the low degree of formalization.

Arctic Council decisions are generally vague and non-binding. Nation states and other relevant actors are merely *encouraged* to follow recommendations from the Council. Therefore, member states are not expected to give Arctic Council decisions precedence over national policies. That said, the amount of power bestowed on the Council by its member states varies by issue area. Most Arctic states have great strategic interests connected to the issue area of oil and gas. The United States and the European Union[14] are net energy importers in need of diversifying their energy supplies and thus have interests in increased hydrocarbon development in the Arctic. Russia's economy is heavily dependent on exports of oil and gas, and the Arctic territories produce 80 per cent of all Russian gas. Norway is also a stakeholder in this, as there are significant oil and gas reserves in the Norwegian parts of the Barents Sea.[15] Further, in Arctic areas, there are unresolved border issues and other jurisdictional issues.[16] These have become even more pressing as hydrocarbons in the region are becoming more accessible. Member states may be willing to accord the Arctic Council some role with respect to environmental aspects related to oil and gas issues – but only to the extent that its work does not interfere with their strategic interests.[17]

The value of preserving the Arctic environment is a basic concept for the Arctic Council. The perceived commitment to taking action to do something about this, on the other hand, varies among member states and according to the interests linked to the issue area in question. An informal rule in the Council is that the country that has taken the initiative to embark on a project is also responsible for follow-up, whereas other countries may contribute to the degree that they find appropriate and to the extent that they find the project interesting and useful. This last point has implications for what we can expect from the Council. As mentioned, the member countries have somewhat differing approaches to what amounts of resources to spend in seeking to mitigate environmental problems in the Arctic. However, it is reasonable to expect that through deliberative processes within the forum and as a result of the value of preserving the Arctic environment, as well as the effort to achieve esteem among the other states, member states will adopt, or at least put their names behind, policies that they would not have approached in the same way in the absence of the Arctic Council.

The Arctic Council's niche

The Arctic Council pays considerable attention to the Arctic marine environment. Petroleum exploitation and transport are identified as the most prominent future threat to the Arctic marine environment, along with POPs.[18] As yet, the Arctic environment has not been affected by oil and gas activities to any considerable degree,[19] but the Council has noted that problems will intensify as Arctic hydrocarbon activities are expected to increase in the future.[20]

The Inari Declaration (Arctic Council 2002) recognizes that the development of oil and gas in many Arctic regions may impact on local standards of living and emphasizes the importance of responsible management of these resources. Included here are emergency prevention and promotion of environmental protection.[21] Air emissions are also identified as potential impacts from oil and gas activities. Finally, the Arctic Climate Impact Assessment (ACIA) study acknowledges that spills are expected and that spill response operations in the Arctic will be more complex and demanding in ice-covered waters than in Prince William Sound or open seas, 'especially since effective response strategies have yet to be developed' (ACIA 2004: 85).

General goals

The issue area of oil and gas within the Arctic Council is linked to policies of sustainable development. This dimension relates to the economic circumstances of indigenous peoples and other residents of the Arctic in the context of preserving the environment (Bloom 1999: 712). A stated goal of the Council is to improve how the Arctic coastal and marine environment is managed, particularly given the accelerated changes due to climate change and increased economic activity (PAME 2004). Measures to prevent oil spills are emphasized, although air emissions from hydrocarbon-related activities are not treated to any considerable degree. The Council's vision for the Arctic marine environment is 'a healthy and productive Arctic Ocean and coasts that support environmental, economic and socio-cultural values for current and future generations' (PAME 2004).

Specific goals and means

Measures taken by the Arctic Council to address the challenges of oil and gas activities in the Arctic are mainly in the form of non-binding guidelines intended to encourage specific policy directions and behaviour – primarily of national authorities, but also of industrial actors. Further, they are meant to contribute to the public's understanding of environmental concerns and practices of Arctic offshore oil and gas activities.[22] Relevant documents produced are the Arctic Offshore Oil and Gas Guidelines (PAME 2002), the Arctic Marine Strategic Plan (PAME 2004) and Field Guide for Oil Spill Response in Arctic Waters (EPPR 1998). Also relevant are the 1997 Environmental Impact Assessment Guidelines (AEPS 1997), although they do not focus directly on oil and gas issues.[23] None of these documents has managed to attract significant attention within Arctic countries. Thus far, most attraction has been directed to the ongoing oil and gas assessment. There is also a shipping assessment being developed by PAME, but progress has been slow.[24] The most specific attempt so far to regulate Arctic oil and gas developments are the oil and gas guidelines.

The Arctic Offshore Oil and Gas Guidelines include all stages of offshore oil and gas activities. The goal is to assist regulators in developing a set of standards, 'which are applied and enforced consistently for all offshore Arctic oil and gas operators' (section 1.2). The Guidelines underline that environmental impact assessment procedures should be used to determine the impacts of offshore oil and gas exploration, development, transportation and infrastructure. States should 'identify and prohibit or restrict oil and gas activities in ecologically and culturally sensitive areas' (section 3.4). Further, it is recommended that 'offshore oil and gas activities should make use of the best available and safest technologies that are determined to be economically feasible and be conducted in a manner to minimize impact on the environment'.[25]

The Oil and Gas Guidelines highlight the existing arrangements in place for coping with environmental problems associated with hydrocarbon activities in the Arctic.[26] Arctic states that are parties to the International Convention on Oil Pollution Preparedness, Response and Cooperation (ORPC 1990) and/or the International Convention for the Prevention of Pollution from Ships (MARPOL 73/78), are required to ensure that operators have oil pollution emergency plans and that these plans are carried out onboard installations.

The guidelines are general in character and only place obligations of *attempt* on the states.[27] The same is true of the Arctic Marine Strategic Plan, which is also neither prescriptive nor binding. As Hayes (2004) puts it: 'it represents the best thinking among national oceans policy-makers in the Arctic – thinking to be shared with all who will listen'.

In sum, the Arctic Council argues that hydrocarbon developments in the Arctic require strict environmental standards because of the sitivity and extreme natural conditions of the environment. Its niche understanding of the problem is special concern for the state of the Arctic marine environment as well as indigenous peoples, in a sustainable development perspective. Measures taken by the Arctic Council to address challenges associated with hydrocarbon activities in the Arctic have consisted mainly of guidelines that pose obligations of attempt on the national states. There are few binding obligations, as important suggestions are modified with expressions as 'may' and 'should' instead of 'shall'. Guidelines from other international regimes are imported, and present and future vulnerabilities are assessed.

Effectiveness

On the basis of the foregoing observations a yardstick of effectiveness will be developed. As Stokke writes in Chapter 2 of this book, 'effectiveness' will be understood in terms of whether the regime has made, or continues to make, a significant contribution to solving the problem it was set up to address (on this, see Keohane *et al.* 1993, Young and Levy 1999). Referring to Underdal (2002), Stokke also notes that effectiveness as problem solving can be measured along three dimensions: output (decisions), outcomes (behaviour) and impacts (on policy objectives).

The output dimension refers to guidelines and agreements adopted by the Arctic Council, with the ultimate aim of improving the state of the Arctic environment. The quality of the initiatives is an important factor. Specific and binding rules of regulation give a positive score on this dimension.

The outcome dimension refers to behavioural change on the national level. Do national decision makers take Arctic Council regulations into consideration when developing their Arctic petroleum policies? A first approach to measuring this is to examine the visibility of the Council in national political debates and public documents, as high visibility will increase the likeliness of behavioural change and thereby the Council's effectiveness potential. Visibility will be measured as number of references to the Council in relevant public documents and news articles covering the issue area.

The Arctic Council itself defines problem mitigation (*the impact dimension*) as the elimination of pollution of the Arctic or improvement of how the Arctic coastal and marine environment is managed. The score on this dimension is highly dependent on the scores on outcome,

on the output dimension. Therefore we will assess
.gating potential on the basis of the two other dimen-
:ping in mind the overall goal of the Council.

ctor relevant for the effectiveness dimension is the degree
ctic states participate in projects through the Arctic Coun-
.y would not else have engaged in. This aspect is included to
gr.   , implications of the institutional character of the Council.

The fact that the issue area of oil and gas involves important national
strategic interests should require specific and binding agreements within the Council, but, as noted, the measures taken have been neither
specific nor binding. Further, when the oil and gas guidelines state that
'offshore oil and gas activities should make use of the best available and
safest technologies that are determined to be economically feasible',[28]
this opens the way for making environmental standards subsidiary to
economic concerns. As environmental considerations may sometimes
impinge on both economic and security interests, we cannot expect
such non-binding guidelines to be followed if they are not compatible
with state interests. The score with regard to output is thus relatively
low. The documents have a certain potential, but this will probably
not be fully realized without further development in terms of their
specificity. As yet there have been no initiatives within the Council for
more specific oil and gas regulations in order to move further from the
general guidelines already developed.

The Norwegian authorities have repeatedly made reference to the
Arctic Council in speeches and documents.[29] For the most part, however, the Arctic Council is mentioned together with other arrangements
in general terms as an arena for international collaboration. A search
through the official websites of the Norwegian government showed
only 12 references to the Council that were found to be oil- and gas-
relevant. Moreover, most of these couple the Arctic Council to the issue
area only loosely, for example by speaking of sustainable development
of resources in general without specifically mentioning oil and gas.[30]
One positive hit was the statement made by environmental minister,
Helen Bjørnøy (2005), to the Norwegian Oil Industry Association,
referring to the Arctic Council as an important collaborative arena
within the case of oil and gas. A slightly greater focus on the Arctic
Council in 2005 may be due to Norway's assuming the chairmanship
in 2006.[31] In any case, Bjørnøy's statement indicates a willingness to
use the Council as an arena for oil- and gas-related issues. However, a
search through specific Arctic Council documents revealed little. For
example, the Oil and Gas Guidelines are mentioned in only one document.[32] Only two references to the Field Guide for Oil Spill Response

[handwritten margin note] ↳ Whereas generic environmental issues may be looked at through the Arctic Council, oil and gas issues tend to be looked at on the national level.

were found, and only three references to the Circumpolar Protected Areas Network (CPAN).[33] Most of the few references to specific Arctic Council work came from the Ministry of Foreign Affairs' presentation of the Arctic Council, which included two of the three references to the Arctic Marine Strategic Plan.

Because of low visibility and references in the form of mere presentations and not active use of the guidelines, potential future effectiveness with regard to outcome must be deemed as low. However, the Norwegian Pollution Control Authority (SFT) did use Arctic Council guidelines actively when providing inputs to the Ministry of the Environment in connection with an opening of the Barents Sea to oil and gas activities,[34] and stressed the importance of an overall environmental impact assessment. This use of Arctic Council requirements to make a statement to the ministry from the SFT represents a kind of self-regulating process within the Norwegian authorities. The fact that Norway is a party to the Arctic Council lends weight to the SFT's arguments, and indicates that the state is regulating its own activities by means of Council decisions.

The Arctic Council is unlikely to have had any noteworthy impact on the process of preparing an environmental impact assessment for the Barents Sea, however. The assessment, finished in 2003,[35] does not fully follow Council guidelines: it is only in line with project-assessment requirements and not strategy requirements, which are more demanding. On the other hand, as one interviewee put it, applying these guidelines is difficult, because they are not very specific.

Further, it should be noted that, as early as in the 1980s, the SFT had been advocating an environmental impact assessment for the Barents Sea. The work of the Arctic Council may thus be seen as one factor that could put further weight behind the SFT's arguments, which were already there in the first place.

From a Norwegian perspective, the biggest current potential for oil and gas lies in bilateral relations with Russia, and not in the Arctic Council. This cooperation has centred on coordinating the regulative framework relating to hydrocarbon activities, especially in terms of bringing Russia's regulative framework in accordance with OSPAR[36] requirements. One interviewee described the bilateral cooperation as being more concrete and engaging Russian institutions directly.[37]

The general public debate, in which the Arctic Council has hardly been noticeable, has largely been defined by environmental NGOs, national regulators and the oil and gas industry. On the other hand, the Council may have played an indirect role here. Particularly through the ACIA, it has managed to put climate change and the particular vulnerability of the Arctic environment on the political agenda, both in Arctic countries and internationally. This in turn has been used by various groups as an argument against hydrocarbon development, and has contributed to the elaboration of a normative framework for general protection of the vulnerable Arctic environment. A link has thus been established between climate and hydrocarbon policies, although this has been done by the wider public and not primarily by the Arctic Council.

In sum, only a few references to Arctic Council documents could be found. The effects of the documents are also difficult to trace, but may in general be perceived as low. Much of the current potential is probably on the bilateral level and not within the Council, making the effectiveness potential in terms of outcome relatively low as well. Thus, the problem-mitigating potential is also assessed as low. However, we should recognize the effect of the Council's role as an agenda setter with regard to concerns for the state of the Arctic environment and with possible, more indirect, implications for problem mitigation.

Other regimes regulating hydrocarbon issues in the Arctic

National domestic laws contain the primary legal controls on the environment, although international environmental laws and principles are playing an increasing role in the Arctic (Nowlan 2001: 11). However, other arrangements already in place seem to provide more promising regulative tools than the Arctic Council. One main initial task of the PAME Working Group was to identify the gaps in coverage by existing international environmental regimes and suggest how these might be filled. PAME's 1996 report concluded that 'there was no urgent need for new international instruments, although some Arctic states had not yet implemented or even ratified existing ones' (Scrivener 1999: 35).

Several global treaties provide legal arrangements applicable to the Arctic. Globally, the most important legal instrument for regulating marine activity is the United Nations Convention on the Law of the Sea (the LOS Convention), which establishes rules governing uses of the oceans and their resources. Seven of the Arctic countries are parties to the LOS Convention, while the USA, though not a signatory, also abides by many of its provisions (Pagnan 2000: 2). In contrast to Arctic Council guidelines, the LOS Convention does not only pose obligations of attempt on states. However, although legally binding, its provisions are quite general.[38]

MARPOL 73/78 is administered by the International Maritime

Organization (IMO) and regulates various types of pollutants, including prevention of pollution by oil (Annex I). Through its marine environment committee the IMO defines areas in need of special protection from maritime activities. Ship-generated pollution has fallen substantially, from approximately 35 per cent of global marine pollution sources in the early 1970s to about 10 per cent by the early 1990s (Nowlan 2001: 21; Rothwell 2000: 60).

The most significant regional initiative to regulate oil and gas issues in the Arctic is the 1992 Convention for the Protection of the Marine Environment of the North East Atlantic (the OSPAR Convention). Although covering only a restricted segment of the circumpolar Arctic, the OSPAR Convention is 'currently one of the most applicable international agreements addressing Arctic marine pollution from various sources' (Arctic Monitoring and Assessment Programme (AMAP) 1998: 2). The Convention covers all technical aspects of pollution from ships of all types.[39] Through the Convention, a strategy[40] has been developed for environmental goals and management mechanism for offshore activities (Muir 2002: 523). Russia is not part of the treaty, although Norway aims at changing this through its bilateral work with Russia.

With so many regimes for dealing with hydrocarbon issues in the Arctic, what role can the Arctic Council play – when its guidelines are general and not binding, when the Council is not visible according to our criteria, and the output and outcome scores thus are low? The next example indicates one gap that the Arctic Council can fill: that of *generating information and knowledge*, with possible impact on other regulative regimes and ultimate problem mitigation. This applies to the work currently undertaken by AMAP with the Arctic Oil and Gas Assessment.

The Arctic Council's niche and its problem-mitigating potential: Knowledge generation

The oil and gas assessment is a follow-up of the oil and gas study in the 1998 AMAP assessment report, which focused on the environmental effects of hydrocarbon activities. The current assessment aims to evaluate four types of effects: socio-economic consequences, environmental impacts from pollution, environmental effects from physical impacts and disturbances; and effects on human health (AMAP 2006). The assessment does not include the relation between hydrocarbon activities and the global CO_2 emissions and greenhouse warming. It is hoped that the assessment will be followed by policy recommendations on

*bon developments in the Arctic, which in turn can form the
or Arctic Council policy recommendations. The target groups
e same as for other Arctic Council projects: politicians and na-
al decision makers as well as the general public.

The USA and Norway co-lead the project steering group, to which
each Arctic Council member state has nominated contributing experts.
The Arctic countries have contributed to the assessment to various degrees. At first, the USA did not want to see the project carried through
and was reluctant to engage in it. According to one of the interviewees,
this changed once socio-economic aspects were included; then the USA
agreed to contribute and even lead this part of the project.[41] The US
participation in the assessment is an example of how the Arctic Council
through its working groups can influence member states to engage in
projects that they would not else have participated in. It is a combination of output (programme activities) and outcome (US participation
and use of resources) and represents an effect of the Council which
ultimately increases its problem-solving potential.[42] One explanation
of the US decision to participate – in addition to the inclusion of socio-
economic aspects – may be the successes of earlier AMAP and ACIA
studies, which may have made it more difficult for Washington to halt
the oil and gas initiative. All in all, the USA contributes considerable
resources by leading the working groups on two chapters. However,
AMAP has experienced difficulties in obtaining necessary data and
information for the project from US experts. The reasons for this are
unclear. It may reflect the central authorities' emphasis on security of
energy supply at the expense of environmental concerns, but the explanation may also lie at the personnel level within the project.[43]

As is the case with most Arctic Council projects, Russia's participation in the oil and gas assessment is sponsored by the other participating states. Lack of Russian data has been a main challenge. Only some
three to four years ago there were usually few problems in obtaining
Russian data, but more recently the tendency within the Ministry of
Natural Resources seems to be, as one interviewee put it, 'to clear
almost every decision and action all the way to the top within the
administration'. Low Russian interest in the environmental challenges
associated with oil and gas development might be another reason.

The European Union does not participate in the oil and gas assessment, and is generally not an active actor within the Council. The
Northern Dimension of the EU focuses on Northwest Russia and the
Baltic Sea states, but little attention has been paid to Arctic Council
work on oil- and gas-related questions. To address the issue of energy
supply, the EU engages in dialogues with Norway and Russia, among

other hydrocarbon-producing countries. Also, the nation states with the EU generally take care of most of their own strategic energy policies, although work is under way to coordinate energy policies on the Union level.[44]

One factor to bear in mind is that Arctic Council member countries send representatives from their environmental and foreign ministries to Council meetings. The oil and gas sector in the member countries has traditionally not been engaged or interested in the Council because of what is perceived as its purely environmental focus. The fact that those who have the closest contact with the oil and gas sector are not represented in the Council may affect its problem-mitigating potential. However, the active participation of the oil and gas industry in the expert meetings of the oil and gas assessment indicates that they perceive the assessment as at least of some importance.

In sum, the oil and gas assessment is facing several important challenges. One is the data input, another is the continued struggles between environmental and industrial and socio-economic interests. It remains to be seen whether the assessment will be a document on which the actors can agree and from which the Arctic states can work out a set of policy principles – and what kind of binding effects these will have. In any case, the assessment is the most important and promising work within the issue area of oil and gas in the Arctic Council. In oil and gas issues, it is almost exclusively this assessment that has been attracting attention among the stakeholders. When the study is released, there will probably be considerable media attention, one way or the other. We may therefore assess the Council's effectiveness potential as being better after this study than before.

Empowerment and region building

Political mobilization/empowerment

Empowerment refers to whether groups that otherwise do not get their voices heard are empowered or mobilized through the Arctic Council (see Chapter 2). Empowerment may be measured according to which groups participate in the Arctic Council and its projects. As indigenous peoples groups are permanent participants in the Council, they have a unique opportunity to make themselves heard on relevant issues like protection of traditional lands and socio-economic aspects. Since indigenous affairs are dealt with in another chapter of this book, let me merely note that there are tensions as regards their role in the Council. In the oil and gas assessment for example, an indigenous peoples'

d a desire to have lead authorship of one of the chapters. did not manage to mobilize enough resources in time to et the deadlines for the overall report, which had already :d by the ministers at the time of their proposal and could ged unilaterally by the assessment leaders. According to one interviewees, indigenous peoples' groups have expressed their dissatisfaction with this. The final draft will show whether they did manage to get their opinions heard.

Lack of resources is also a problem facing NGOs. The Worldwide Fund for Nature (WWF) for example, has observer status in the Arctic Council. This has obvious advantages in terms of increased access to stakeholders and information at a high level, but there is also the question of whether the costs of participation are worth the results. Also, in cases of disagreement as to the content of Council projects, as one interviewee put it, 'you get a tension between what is right vis-à-vis insider status and what is right vis-à-vis the role as an environmental and campaigning organization.'

As regards the oil and gas assessment, the meetings of assessment authors and experts have been attended by participants from indigenous peoples' organizations and NGOs, who have also reviewed drafts and commented on them. Thus, indigenous peoples and environmental NGOs may make themselves heard in the process, but a conclusion on this point will have to wait until the study is published and we can see to what degree the interests of these and other groups have been taken into consideration. The real problem is the Arctic Council's low visibility in general and particularly on oil and gas issues. Thus, to be heard within the Arctic Council is not the same as being heard on the international arena or even on national arenas within Arctic countries.

In addition to including various groups in its activities, the Arctic Council may also serve as a provider of arguments for various other groups concerned with the state of the Arctic environment, groups not closely linked to the Council. This may also be measured according to visibility. We should, for example, expect environmental NGOs to rely on all possible relevant measures, including Arctic Council guidelines and studies where relevant, in trying to impact on national authorities in the debate over oil and gas development in the Arctic.

The Arctic Council is hardly found in the vocabulary of Greenpeace and Bellona in the debate over Arctic oil and gas issues.[45] One explanation may be that the Arctic Council has not to any considerable degree addressed the question of whether such development should be taking place at all. It may also be an indication of a lack of faith in the Arctic Council's potential as a regulator within the field. That said, through the generation of better information, and through the 2004 ACIA report in particular, climate change has contributed to putting Arctic oil and gas issues on the international agenda.

WWF as observers to the Arctic Council do not refer more often to the Council in oil and gas issues in mass media than do Bellona or Greenpeace.[46] However, on the WWF International home pages we find many references to the Council in relation to hydrocarbon issues, and the Arctic Council is identified as 'an important policy arena for the WWF'.[47] Thus, WWF uses the Arctic Council in its work with oil and gas issues – without, however, referring to the organization in the media. If the Arctic Council were a politically important body, it would probably be referred to more frequently by the various stakeholders. Another explanation may be that the vagueness and non-binding character of what has been produced by the Council makes reference difficult.

Region building

Region building refers to whether the Arctic Council contributes to an elevation of the Arctic as a distinct region on the international political agenda. With regard to discursive regionality as outlined by Stokke in Chapter 2, visibility is again a key word: the Council must be 'visible' and 'present' in the minds of Arctic residents. I have already argued that environmental NGOs do not use Arctic Council work as backup for their arguments, but is the regional press nevertheless conscious of the Council's work? The answer again seems to be negative. In the *Anchorage Daily News* we find that the Arctic Council is mentioned 17 times between 1996 and 2006 together with oil and gas issues, but in most of these the direct link to Arctic Council work is the ACIA study on Arctic warming. Only three times was the Arctic Council mentioned in direct relation to oil and gas issues.[48] The northern regional Norwegian newspaper, *Nordlys*, shows 32 hits for Arctic Council and oil/gas/petroleum from 1996 to 2006, but only in nine of these is the Council mentioned in relation to hydrocarbon issues.[49] In comparison, a search for *nordområdene* (lit: 'the northern areas'), the Norwegian word for the northern parts of Norway and the connected sea areas, and otherwise using the same combination of words, showed 361 hits from *Nordlys*. A search through the whole database yielded 2,206 hits when 'Arctic Council' was replaced by the term *nordområdene*.

Moreover, a study under the Finnish chairmanship of 2000–02 concluded that the Arctic Council was 'rather poorly known among the inhabitants of the Arctic countries' (Turunen and Kankaanpää 2002: 4).

Public officials were included among the respondents. The fact that PAME and EPPR (the working groups that, before the oil and gas assessment, had been most involved in hydrocarbon-related work), were among the least-known indicates a fairly low region-building potential within this issue area. By contrast, AMAP stood out as the best-known working group, which indicates that the oil and gas assessment could have some potential for contributing to region building. On the other hand, we may get a situation where the assessment receives a lot of attention – and not the Arctic Council. The Arctic Council is 'intrinsically not interesting for the public', as one interviewee pointed out. The Arctic as such is very interesting – but not the Arctic Council. If we look at how the ACIA has been presented in the media, we see that the Arctic Council, if mentioned, is not highlighted at all, but is merely a footnote.

Thus, the oil and gas assessment may become a contributing factor, but as yet the Arctic Council does not play a significant role as a discursive region builder within the issue area of oil and gas. On the other hand, the issue area in itself – the economic aspect in particular –may prove to be a strong region-building factor. Here a relevant question is whether the Arctic may really be considered a region as such, or whether we should confine ourselves to talking about for instance the 'Eurasian Arctic' and the 'North American Arctic'. The increased interest in the Arctic region in recent years must to a great extent be accredited to the renewed optimism around the oil and gas sector, particularly in the Eurasian Arctic – perhaps most of all Russia's gigantic Shtokman gas field.[50] In Norway, there is political will to invest resources in the Arctic Council in general, hydrocarbon issues included. The assumption is that Norway, through collaboration with other states, can secure its strategic interests related to natural resources, jurisdictional and security issues in the High North. On this point, it is not the Arctic that is the Norwegians' reference when speaking of hydrocarbon issues in Northern areas, but rather Norway in relation to, and sometimes in opposition to, other Arctic states as well as the European Union. This may change as the Arctic Council and this particular issue area mature, but one constraining factor is again that hydrocarbons are closely linked to national strategic interests. Another constraining factor, and one that cuts across all issue areas, is that the Arctic states are not 'purely Arctic', since their centres of gravity are located in more southern areas. Finally, much of the work within the organization has focused on mitigating environmental problems located in and originating from Northwest Russia. This makes it more difficult to maintain a truly circumpolar focus, which again impacts negatively on the potential for constructing an Arctic identity.

Conclusion

The low degree of formalization, in combination with relatively non-binding agreements and a strategically important issue area, makes it unrealistic to believe that ambitious goals, like that of eliminating pollution of the Arctic, will be an immediate result of the Arctic Council's work. The Council does not serve as an important forum for the great powers in their energy policies. At this stage, other issues linked to security of energy supply – such as attractive investment climates, infrastructure and security of transport – are more likely to win great-power attention because of their strategic importance. Further, the issue area of oil and gas is regulated through various global and regional regimes, some of which are more specific and binding than Arctic Council agreements. Since Arctic Council guidelines are based largely on these, it is difficult to track which changes are the effects of Arctic Council agreements and which may be accredited to other regulative mechanisms, although it seems likely that other arrangements should be accredited the regulative role. The effects of the Council's work do not yet seem coherent with its aspirations to 'improve how the Arctic coastal and marine environment is managed, particularly given the accelerated changes occurring in the north due to climate change and increased economic activity' (PAME 2004). The main challenge facing the Arctic states as regards the oil and gas sector and related environmental challenges is to determine on specific appropriate environmental constraints for petroleum-related activities in the area and to provide structures for improved adherence to such constraints. We must conclude that the Council scores low on effectiveness as a whole. The greatest potential lies, not in the regulative role, but in generating better information and knowledge about the effects of hydrocarbon activities, as well as in the agenda-setting role. The issue area of oil and gas has not yet been a main priority of the Council, but is 'up and coming' within the organization, so we may conclude that this potential has not yet been fully realized.

Low empowerment is indicated by the low number of references to the Arctic Council as well as its limited role on the international and national arenas. However, through the Council, both environmental NGOs and indigenous peoples groups have gained access to stakeholders and information that they would not else have had. In this sense, the empowerment score is positive. The Arctic Council does not seem to play an important role as regards region building within the issue area. The issue area in itself, on the other hand, may have a significant region-building potential, although the question remains whether the Arctic should be considered as a region in its own right, or whether it is composed of different regions, such as the Eurasian or the North American Arctic.

Lastly, the need for esteem and the need to promote the shared value of preserving the Arctic environment give the Council a role to play, enabling states to impact on each other and participate in environmental projects they would not else have engaged in. In this sense, the Council's problem mitigation potential is positive compared to what the situation would have been without the Council. The AMAP oil and gas assessment is one example of such a process. The question is whether the result of the assessment will be policy recommendations and important political commitments. It also remains to be seen whether the Council is now at a stage of institutional development and will develop into a regulative body within Arctic oil and gas issues, or whether it will find its role as a knowledge generator. Much will depend on the forthcoming oil and gas assessment and its political implications. Given current constraints, an oil and gas assessment that could produce sound policy recommendations and mobilize public attention may be all we should expect from the Arctic Council at this stage.

Notes

1 BEAR is currently involved in an analysis that focuses mainly on socioeconomic effects of hydrocarbon developments on the northernmost counties of Norway; environmental issues have not to any considerable degree been addressed by BEAR in assessments or policy documents.
2 The interviewees are listed in the reference list, but are not directly referred to by name in the text.
3 For a more detailed discussion of Arctic oil and gas reserves, see Sawhill (2005).
4 See, for example, *Petroleum Intelligence Weekly* (2004: 6).
5 In December 2005, however, the last of a series of proposals to open the Arctic National Wildlife Refuge in Alaska to oil drilling was blocked by the US Senate.
6 In 2002, 80 per cent of Russia's gas was produced at fields with declining production. Significant new capacity will have to come on stream over the next two decades if current rates of production are to be maintained (Kim 2005: 366).
7 In Norway, for instance, there was a similar debate both in the early 1970s and the 1980s. Many people expected oil and gas activities in the Barents Sea to escalate then.

8   In the Canadian Arctic, tankers load from the shore via hoses on Cameron Island. There is no tanker transport of crude oil in the US Arctic, as North Slope production is transported via the Trans-Alaska pipeline (PAME 2000: 4).

9   Another important bottleneck are the high insurance costs of operating in these waters.

10   Pipeline accidents are also a significant source of pollution.

11   For example, oil from the 'Exxon Valdez' oil tanker accident in 1989 was still embedded in Prince William Sound beaches in the summer of 2003 (ACIA 2004:85).

12   Low species diversity and biological productivity and long-lived organisms with high lipid levels are among the factors that contribute to the sensitivity of the Arctic ecosystem (AEPS 1991: 9).

13   Two poles in the debate are environmental NGOs and industrial actors like oil and gas companies. There may also be tensions between various sectors of national governments, e.g. the environmental and petroleum ministries.

14   The EU is only an ad hoc observer in the Arctic Council.

15   Although estimates show that the greatest potential is in the Russian Arctic.

16   The USA and Canada have not yet agreed upon a common maritime boundary in the Beaufort Sea. Norway and Russia face a similar problem over their boundaries in the Barents Sea. Moreover, there is a dispute between the contracting parties to the 1920 Svalbard Treaty as to whether Norway can claim exclusive economic zone and continental shelf zones around Svalbard, thereby excluding the economic rights guaranteed to the other contracting states through the Treaty.

17   Military issues were for example explicitly left out of the Arctic Council mandate in the Ottawa Declaration.

18   PAME (2002, §1.4).

19   AMAP (1998: 693) concluded: 'With the exception of the concentrations in a few areas exposed to catastrophic releases from oil pipeline ruptures or prolonged chronic releases, hydrocarbon levels in the Arctic associated with anthropogenic inputs have been found to be relatively low and not of ecological consequence.'

20   In the years after the AMAP report was published, oil transportation in Arctic waters has, as noted earlier, increased dramatically.

21   The Council also highlights that particular attention must be paid to the impact of development and use of natural resources on indigenous peoples and their communities (Arctic Council 2002).

22   See, for example, PAME (2002, §1,2).

23   See bibliography for Internet addresses with full texts of the various guidelines.

24   For more information about this, see www.pame.is

25   Concerning standards of platforms and other installations, the International Organization for Standardization (ISO) is mentioned. Further, Det Norske Veritas and the American Petroleum Institute are mentioned as having recommended practices for design of offshore Arctic pipelines (section 6.3).

26   (Section 6.3) The International Organization for Standardization (ISO

14000 series), the American Petroleum Institute (recommended Practice 75) and Oil and Gas Producers (OGP) and UNEP/OGP publications for certification and MARPOL 73/78 are all mentioned, in varying contexts.

27 In annex F of the Guidelines, even less committing measures are presented in a list of elements that are described as 'elements that *may* be incorporated into company safety and environmental policies and objectives' (author's emphasis).

28 PAME (2002, section 6.3) (see above section).

29 A search through www.odin.dep.no, official informational website of the Norwegian government and ministries, yielded 325 hits for 'Arctic Council'.

30 Only four of the 12 'positive' hits were identified as linking the Arctic Council directly to oil and gas issues.

31 It may also be a result of what one interviewee described as a greater focus on the maritime environment on the part of the Norwegian authorities in recent years.

32 A search was made for the period from 1 January 1998 to 18 November 2005.

33 Little political will to follow up CPAN contributes to the impression that the Arctic Council does not advocate an Arctic Ocean without hydrocarbon developments to any considerable degree, but emphasizes the importance of strict environmental standards in view of increased developments. See www.arctic-council.org for further details about CPAN.

34 'Konsekvensutredning for utbygging og drift av Snøhvit LNG. Miljødirektoratets vurdering av konsekvensutredningen' and 'Petroleumsvirksomheten i Barentshavet' (available at www.sft.no).

35 'Utredning av konsekvenser av helårlig petroleumsvirksomhet i området Lofoten – Barentshavet'. For more details see the home pages of the Norwegian Ministry of Petroleum and Energy.

36 Convention for the Protection of the Marine Environment of the North East Atlantic. For full text of the Convention, see http://www.ospar.org/eng/html/convention/welcome.html (accessed 10 December 2005).

37 It is also worth mentioning that the Russian legal framework is already quite strict, with requirements of zero discharge. The problem, however, is implementation. For example, according to one interviewee, Russian authorities tend to follow up foreign companies closely, whereas Russian companies do not appear to be controlled to the same extent.

38 Article 192 of the Convention reads as follows: 'States have the obligation to protect and preserve the marine environment.' Further, 'states shall take all measures necessary to ensure that activities under their jurisdiction or control are so conducted as not to cause damage by pollution to other States and their environment, and that pollution arising from incidents or activities under their jurisdiction or control does not spread beyond the areas where they exercise sovereign rights in accordance with this Convention' (Article 194.2).

39 Particular types of ships are not specified. The Convention speaks of vessels in general: see, for example, article 10.

40 Offshore Oil and Gas Industry Strategy, see http://www.ospar.org/eng/html/sap/welcome.html#offshore (accessed 10 December 2005).

41 A chapter on past, present and future activities is also US-led.

42 Evidently, the USA also affects the project through its participation, first and foremost by also including socio-economic aspects in the analysis.

43 US reluctance has also been an issue in previous AMAP reports. One explanation may be that environmental politics in the Arctic, or the Arctic itself, is not a major topic for the USA. It remains to be seen whether the increased focus on hydrocarbon developments in the region will change this.

44 Based on informal conversations with several bureaucrats within the EU-system, Brussels and Oslo, autumn 2005 and spring 2006.

45 A search was made through Norwegian media from February 1996 to March 2006, using the Atekst database. Two hits were found for Bellona* AND 'Arctic Council' AND (oil* OR gas* OR petroleum*). In comparison, Bellona* AND nordområde* AND (oil* OR gas* OR petroleum*) showed 122 hits. Only one hit was found for Greenpeace AND 'Arctic Council', and this was not hydrocarbon relevant. In comparison, Greenpeace AND Arctic* yielded 83 hits.

46 For the same period, WWF mentioned the Arctic Council and oil or gas together only three times. The Arctic Council was mentioned eight times.

47 See for example: http://www.panda.org/about_wwf/where_we_work/arctic/projects/index.cfm?uProjectID=9E0077 (accessed 15 February 2006).

48 A search was made for the combination of 'Arctic Council' AND (oil* OR gas* OR petroleum*). Climate change was used as an argument against hydrocarbon developments in the Arctic in two articles. A search for only 'Arctic Council' gave 21 hits.

49 The search was made in the Atekst data base of Norwegian newspapers.

50 Archer views hydrocarbon developments in the Arctic as a contributing factor in the process of change from the Arctic as a region based on a security conflict frame to that of increased international cooperation, particularly within scientific research and environmental protection. 'While petroleum exploration has not been *the* major impetus for either activity, it has provided a stimulus' (Archer 1992: 98, emphasis in original; Fogelson 1992: 166; Keskitalo 2004: 35–6; Nassichuk 1987; Osherenko and Young 1989; Roots 1992).

References

Archer, C. (1992) 'Arctic Cooperation', in J. Käkönen (ed.), *Vulnerable Arctic: Need for an Alternative Reorientation*, Tampere Peace Research Institute, Research Report no. 47.

Arctic Climate Impact Assessment (ACIA) (2004) *Impacts of a Warming Arctic*, Cambridge: Cambridge University Press.

Arctic Environmental Protection Strategy (AEPS) (1991) *Declaration on the Protection of Arctic Environment*, Online. Available HTTP: <http://www.arctic-council.org/files/infopage/74/artic_environment.pdf>.

AEPS (1997) Guidelines for *Environmental Impact Assessment in the Arctic*.

Arctic Monitoring and Assessment Programme (AMAP) (1998) *Assessment Report: Arctic Pollution Issues*, Oslo; AMAP.

AMAP (2006) *Assessment of Oil and Gas Activities in the Arctic – Process and Outline Content*, Online. Available HTTP: <http://www.amap.no/MiscTempFiles/OGA%20Outline%20-%20January%202006.doc> (accessed 17 January 2006).

Arctic Council (1996) *Declaration on the Establishment of the Arctic Council* (The Ottawa Declaration), Online. Available HTTP: <http://www.arctic-council.org/en/main/infopage/190/> (accessed 10 November 2005).

Arctic Council (2000) *Barrow Declaration on the Occasion of the Second Ministerial Meeting of the Arctic Council*, Online. Available HTTP: <http://www.arctic-council.org/files/infopage/75/bar_decl.pdf> (accessed 10 November 2005).

Arctic Council (2002) *Inari Declaration on the Occasion of the Third Ministerial Meeting of The Arctic Council*, Online. Available HTTP: <http://www.arctic-council.org/files/infopage/75/inari_Declaration.pdf> (accessed 10 November 2005).

Årøy, Y. (2005) 'Oil transportation from Northwest Russia and Norwegian Coast Emergency Response System', in A. Bambulyak and B. Frantzen (eds.), *Oil Transport from the Russian Part of the Barents Region*. Svanhovd Environmental Centre, Svanhovd.

Bjørnøy, H. (2005) *Miljøutfordringene i nordområdene*, speech at the Norwegian Oil Industry Association's annual conference, SAS Plaza Hotel, Oslo, 10 November 2005, Online. Available HTTP: <http://www.olf.no/?28794.pdf> (accessed 7 January 2006).

Bloom, E. T. (1999) 'Current developments: establishment of the Arctic Council', *American Journal of International Law*, 93 (3): 712–22.

Convention for the Protection of the Marine Environment of the North East Atlantic (the OSPAR Convention), Online. Available HTTP: <http://www.ospar.org/eng/html/convention/welcome.html> (accessed 10 December 2005).

EIA (2006) *World Proved Reserves of Oil and Natural Gas, Most Recent Estimates*, Online. Available HTTP: <http://www.eia.doe.gov/emeu/international/reserves.html> (accessed 30 January 2006).

Emergency Prevention, Preparedness and Response, EPPR (1998) *Field Guide for Oil Spill Response in Arctic Waters*, Environment Canada, Yellowknife, NT Canada, Online. Available HTTP: <http://eppr.arctic-council.org/fldguide/fldguide.pdf> (accessed 10 November 2005).

Fogelson, N. (1992) *Arctic Exploration & International Relations 1900–1932: A Period of Expanding National Interests*, Fairbanks, AK: University of Alaska Press.

Hayes, M. F. (2004) *Current Oceans Policy: United States Perspectives*, Remarks at the Conference of International Energy Policy, St. Petersburg, Russian Federation, 24 June 2004, Online. Available HTTP: <http://www.state.gov/g/oes/rls/rm/2004/34146.htm> (accessed 10 April 2006).

IMO (1993) 'Impact of Oil and Related Chemicals and Wastes on the Marine Environment', GESAMP Report No. 50, London.

International Convention for the Prevention of Pollution from Ships (MAR-POL 73/78), Online. Available HTTP: <http://www.imo.org/Conventions/contents.asp?doc_id=678&topic_id=258> (accessed 9 April 2006).

International Convention on Oil Pollution Preparedness, Response and Cooperation (ORPC) (1990), more info at: <http://www.imo.org/Conventions/contents.asp?topic_id=258&doc_id=682>

Jumppanen, P. (1990) 'Environmental aspects of the exploitation of oil and gas reserves', in L. Heininen (ed.), *Arctic Environmental Problems*, Tampere Peace Research Institute, Occasional Papers No. 41.

Keohane, R. O., Haas, P. M. and Levy, M. A. (1993) 'The effectiveness of international environmental institutions', in P. M. Haas, M. A. Levy and R. O. Keohane (eds.), *Institutions for the Earth: Sources of Effective International Environmental Protection*, Cambridge, MA: MIT University Press.

Keskitalo, E. C. H. (2004) *Negotiating the Arctic: The Construction of an International Region*, London: Routledge.

Kim, M. (2005) 'Russian oil and gas:impacts on global supplies' *Autraliancommodities* 12, 2: 361–78, Online. Available HTTP: <http://abareonlineshop.com/PdfFiles/PC13159.pdf>

Kitsos, T. (2000) 'Setting environmental research priorities', paper presented at the National Association of State Universities and Land Grant Colleges, Section on Mineral and Energy Resources, September 2000, Online. Available HTTP: <http://www.mms.gov/ooc/newweb/speeches/serp.htm> (accessed 10 November 2005).

Muir, M. A. K. (2002) 'Integrated coastal marine management in northern regions: reconciling economic development and conservation', *Journal of Coastal Research*, Special Issue 36.

Nassichuk, W. W. (1987) 'Forty years of Northern non-renewable natural resource development', *Arctic*, 40: 274–84.

Nowlan, L. (2001) *Arctic legal regime for environmental protection/The World Conservation Union – IUCN*, Environmental Policy and Law Paper no. 44, Bonn: Environmental Law Centre.

OECD/IEA (2004) *World Energy Outlook*, Paris.

Offshore Oil and Gas Industry Strategy, Online. Available HTTP: <http://www.ospar.org/eng/html/sap/welcome.html#offshore> (accessed 10 December 2005).

Osherenko, G. and Young O. (1989) *The Age of the Arctic: Hot Conflicts and Cold Realities*, Cambridge: Cambridge University Press.

Pagnan, J. L. (2000) 'Arctic Marine Protection' *Infonorth* 53 (4): 469–76.

Protection of the Arctic Marine Environment (PAME) (1996) PAME Report to the Third Ministerial Conference on the Protection of the Arctic Environment, 20–21 March.

—— (2000) 'PAME – Snapshot Analysis of maritime activities in the Arctic', Online. Available HTTP: <http://www.pame.is/sidur/uploads/Norwegian_maritime_directorate.PDF> (accessed 18 April 2006).

—— (2002) 'Arctic Offshore Oil and Gas Guidelines', Online. Available HTTP: <http://www.pame.is/sidur/uploads/ArcticGuidelines.pdf>

—— (2004) 'Arctic Marine Strategic Plan', Online. Available HTTP: <http://www.pame.is/sidur/uploads/PAME%20Bæklingur%20ens%20-%20net.pdf> (accessed 1 November 2005).

Petroleum Intelligence Weekly (2004) October 4.

Ragner, C. L. (2000) 'The Northern Sea Route – commercial potential, economic significance, and infrastructure requirements', *Post-Soviet Geography and Economics*, 41 (8): 541–80.

Roots, E. F. (1992) 'Cooperation in Arctic science: background and requirements', in F. Griffiths (ed.), *Arctic Alternatives: Civility or Militarism in the Circumpolar North*. Canadian Papers in Peace Studies, No. 3. Science for Peace/Samuel Stevens, Toronto.

Rothwell, D. R. (2000) 'Global environmental protection instruments', in D. Vidas (ed.), *Protecting the Polar Marine Environment – Law and Policy for Pollution Prevention*, Cambridge: Cambridge University Press.

Sawhill, S. (2005) 'Ressurser i nordområdene – hva vet vi?' *Horisont*, 3.

Scrivener, D. (1999) 'Cirumpolar cooperation to protect the Arctic environment', *Journal of Maritime Education*, 15 (3): 33–38.

Statens Forurensningstilsyn, SFT (2001) *Konsekvensutredning for utbygging og drift av Snøhvit LNG. Miljødirektoratets vurdering av konsekvensutredningen*, Online. Available HTTP: <http://www.sft.no/nyheter/brev/snohvit_konsekvensutredning200601.htm> (accessed 9 April 2006).

Statens Forurensningstilsyn, SFT (2001) *Petroleumsvirksomhet i Barentshavet*. Online. Available HTTP: <http://www.sft.no/nyheter/brev/barentshavetbekymringsbrev080801.htm> (accessed 9 April 2006).

Turunen, M. and P. Kankaanpää (2002) *The Visibility of the Arctic Council – Results of a Background Study*, Arctic Centre Report No 37, Helsinki: Edita Prima.

Underdal, A. (2002) 'One question, two answers', in E. L. Miles, A. Underdal, S. Andresen, J. Wettestad, J. B. Skjærseth and E. M. Carlin, *Environmental Regime Effectiveness: Confronting Theory with Evidence*, Cambridge: MIT Press.

United Nations Convention on the Law of the Sea (1982) Online. Available HTTP: <http://www.un.org/Depts/los/convention_agreements/texts/unclos/closindx.htm>

United States Geological Survey (2000) *World Petroleum Assessment*, Online. Available HTTP: <http://pubs.usgs.gov/dds/dds-060/> (accessed 9 April 2006).

Young, O. (2000) 'The structure of Arctic cooperation: solving problems/seizing opportunities', paper prepared at the request of Finland in preparation for the Fourth Conference of Parliamentarians of the Arctic Region, Rovaniemi, 27–29 August.

Young, O. and Levy, M. A. (1999) 'The effectiveness of international regimes', in O. R. Young (ed.), *The Effectiveness of International Environmental Re-*

gimes: Causal Connections and Behavioral Mechanisms, Cambridge, MA: MIT Press.

Interviews:

Døvle, P. (2005) 6 December, SFT, Oslo
Kammerud, A. (2006) 14 March, SFT, Oslo.
Reiersen, L. (2005) 30 November, AMAP Secretariat, Oslo.
Rosenberg, S. (2005) 24 November, Norwegian Ministry of Foreign Affairs, Oslo.
Smith, S. (2006) 3 March, WWF, Oslo.
Syvertsen, E. (2005) 6 December, SFT, Oslo.

8 International institutions and Arctic governance

Olav Schram Stokke

Introduction

This book has examined the consequences of Arctic institutions on important socio-economic issues related to indigenous concerns, communicable disease control, pollution and conservation, climate change, as well as oil and gas activities.[1] Applying the analytical framework presented in Chapter 2, several case studies have examined the impacts in those issue areas with regard to three circumpolar or sub-regional initiatives that emerged due to improvements in East–West relations two decades ago: the Arctic Council, the Barents Euro-Arctic Region, and the Baltic Sea Region. Institutional consequences have been studied at three levels: (1) effectiveness, defined as mitigation or removal of specific problems addressed by the institution; (2) political mobilization, highlighting changes in the pattern of involvement and influence in decision making on Arctic affairs; and (3) region-building, understood as contributions by Arctic institutions to denser interactive or discursive connectedness among inhabitants in the region. Those types of impact structure this concluding chapter, which summarizes and compares findings from the five in-depth case studies presented in preceding chapters.[2]

Institutional effectiveness and niche selection

The 'effectiveness' of an international institution refers to its ability to contribute significantly to removing or mitigating the problem that motivated its formation. In Chapter 2, we saw that, in the case of Arctic institutions, such effectiveness must be addressed with keen attention to their interplay with the broader set of institutions existing at global or regional levels that already address closely related matters. This is partly because Arctic environmental and socio-economic challenges are entwined with economic activities in other parts of the globe; but also among Arctic institutions, interplay is relevant since these are bodies that cover partly overlapping issue areas and emerged somewhat haphazardly in the course of only a few years.

The 'niche' concept was introduced in Chapter 2 to capture an institution's specialization as to what part of the problem it deals with and the type of resources it brings to bear on problem solving – and how such specialization relates to activities under other international bodies. I outlined three such niches, based on whether the institution in question specializes in cognitive activities (i.e. seeks to improve the level of knowledge about regional problems), in strengthening norms relevant to those problems, or in enhancing regional capacity to cope with them.

Niche selection revolves around whether the institution concentrates on aspects of problem solving that it is particularly well equipped or positioned to cope with. In this respect, at least three distinctive features of Arctic institutions are relevant. First, they were all set up in response to the cooperative window of opportunity that arose in the late 1980s with the improvement in East–West relations, and were largely motivated by a desire to involve Russia in robust cooperative structures and thereby enhance regional security. Second, Arctic institutions typically address a wide range of issue areas, including environmental protection, commerce and industry, health, education, and cultural affairs. A third feature that sets Arctic institutions apart from many others is participatory heterogeneity. In Chapter 1, we showed that these institutions tend to involve representatives not only of national governments but also of provincial governments, indigenous organizations where appropriate, and other civil society groups.

Taken together, the long-term security objectives and the broad functional scope have generated a willingness among Western participant states to pay a disproportionate share of the costs of Arctic collaborative endeavours. Especially at the sub-regional level, where economic and environmental interdependencies are the strongest, such willingness has also extended to costly capacity-enhancement programmes undertaken in Northwest Russia and the Baltic states. For its part, the participatory heterogeneity of Arctic institutions implies that actors who are marginal in other international arenas relevant to the North have come to see Arctic institutions as promising vehicles for pursuing their political goals. As discussed below, participatory heterogeneity is important for political empowerment and the building of a stronger regional identity, but it also impinges upon the ability of

Arctic institutions to induce changes that can contribute to solving the various problems they address.

Environmental research and monitoring

Grave environmental challenges, not least in the Russian North, were among the driving forces for establishing the institutions examined here. All three institutions are characterized by functional breadth, but activities ultimately aimed at alleviating environmental problems were among the first practical outcomes. This was in part because the new institutions were able to build upon pre-existing structures, notably dyads of bilateral environmental agreements drawn up in the late 1980s between the Soviet Union and most other Arctic states (Stokke 1990). Later on, when the Arctic Council was formally established, it subsumed within its purview the Arctic Environmental Protection Strategy (AEPS), a mechanism for initiating and coordinating monitoring, normative and capacity-enhancement activities that already had a substantial track record.[3]

While the sub-regional bodies for Arctic environmental governance have focused on capacity enhancement, the Arctic Council has mainly opted for a niche that is cognitive – notably in terms of its environmental monitoring activities, which have emerged as the 'specialization of the Arctic Council' (Stenlund 2002: 837). Its Arctic Monitoring and Assessment Programme (AMAP) examines pathways and levels of hazardous contaminants, including persistent organic pollutants (POPs), heavy metals, radionucleides and hydrocarbons; examines their effects on human health and Arctic flora and fauna; and assesses impacts of climate change (AMAP 2004). Other permanent working groups under the Council focus on protection of the marine environment, emergency prevention, preparedness and control and conservation of Arctic flora and fauna. As noted in Chapter 5, these various working groups have produced a series of high-profiled reports on Arctic challenges, including two comprehensive AMAP Assessment Reports and several more specific ones in areas like as health and oil/gas, with an Arctic Marine Shipping Assessment due in 2008.[4]

The fact-finding endeavour to attract the greatest external attention so far is the Arctic Climate Impact Assessment (ACIA), coordinated jointly by the Arctic Council and the International Arctic Science Committee. As discussed by Hoel in Chapter 6, that assessment has aimed at taking stock of existing knowledge about the regional consequences of climate change. In the end, substantial new knowledge had to be generated as well, clarifying the diversity of impacts across the Arctic region.

It was already known that temperature rises in polar areas would be roughly twice the global average, with potentially strong effects on the heat exchange between land, air and water. The ACIA has spelt out in greater detail the impacts of a distinctly Arctic feedback mechanism, by which receding snow and ice will boost heat absorption and accelerate further melting. Tree-lines are expected to move hundreds of kilometres northwards; shifts in the occurrence of marine and terrestrial living resources are among the consequences of climate change that will affect Arctic residents directly. The dual role assumed by the USA in the global climate regime – rejecting commitments to specific emissions targets but contributing heavily to the scientific assessment work – is evident also here. The ACIA secretariat was placed at the University of Alaska in Fairbanks, and the Assessment Steering Committee was headed by a US scientist.

Specializing in environmental monitoring allows the Arctic Council to take advantage of its circumpolar scope and the priority given by its member states to polar science. Involving scientists from all Arctic eight states enables more efficient compilation of environmental and human health data from the entire region than would be possible under bilateral or sub-regional structures. Although the Arctic region is heterogeneous in many respects, it has important commonalities like slow regenerative capacity, certain hot-spots highly vulnerable to disturbances, and joint placement at the receiving end of long-distance contaminant flows (AMAP 2002). At the same time, environmental monitoring is in the interest of all states involved – and, at least in the short run, it does not raise questions that are controversial in Arctic societies or between states, as regulatory issues might do. As such, a niche strategy focused on cognitive rather than normative contributions exploits the particular edge that the Arctic Council has over other bodies, without challenging the interests of any Arctic states.

Protecting the Arctic environment

When it comes to *regulating* the activities that give rise to Arctic environmental problems, Arctic institutions have made far more modest substantive contributions than those in the cognitive, fact-finding domain. Having examined the various types of threats and the scope and strength of international legal instruments on the marine environment, the working group on the matter concluded that there was no compelling need for new binding Arctic-level instruments in this area (PAME 1996). Instead, as noted in Chapter 5, non-binding and relatively general guidelines have been elaborated for use by national agencies, such as the Offshore Arctic Oil and Gas Guidelines. The standards contained here are derived from and invoke existing and legally binding instruments, including the Law of the Sea Convention, various agreements drawn up under the International Maritime Organization, and regional conventions. Such normative nesting – adaptive and subordinate placement within a broader institution – may contribute to environmental problem solving *if* the Arctic application adds specificity or determinacy to existing commitments, or *if* it engages Arctic states that have failed to join the broader instrument.[5] This is hardly the case for the Oil and Gas Guidelines, given their soft-law nature and low visibility in the relevant bureaucracies. Both for that document, and the more specific guidance provided by Arctic Council working groups on matters like oil transfer, oil-spills response and conservation of certain sea birds, the non-bindingness and lack of systematic review of implementation are likely to limit any impacts on the policies of Arctic states, province-level authorities and vessel operators.[6] Also the Barents Euro-Arctic Region (BEAR) cooperation has maintained a very low profile with respect to regulating activities that lead to marine pollution.[7]

In some cases, Arctic institutions have played a role in what we may term a 'catalytic approach' to regulation.[8] The most prominent instances relate to the documentation provided under AMAP, showing how POPs bio-accumulate in the fatty tissue of fish and mammals that figure prominently in the diet of many Arctic residents. Thanks to close interaction between AMAP and the POPs Task Force under the Convention on Long-Range Transported Air Pollution (CLRTAP), those findings were fed into the process of negotiating the 1998 Århus Protocols on POPs and heavy metals under CLRTAP. The Arctic Council also provided a platform whereby Inuit organizations and other Northern players could strengthen their relations to foreign policy makers – especially in Canada (Fenge 2003) but also in the USA (Huntington and Sparck 2003) – and emphasize the Northern significance of the POPs issue. All considered, Arctic Council activities did contribute – albeit not decisively – to the regulatory clout of the 1998 Århus Protocols and the global Stockholm POPs Convention adopted three years later.[9]

While it should not be overstated, the influence of Arctic-Council based activities on broader regulatory strengthening on POPs was largely based on the environmental monitoring programme developed under this particular institution. It also resulted from another distinctive feature of the Council – the unusually prominent place given to representatives of indigenous peoples. As Permanent Participants, Inuit organizations obtained direct and regular access to high-level officials in the foreign ministries of Arctic states – and biannually even to ministers. Those two features – coordination of an extensive and widely recognized research and monitoring endeavour, and participatory heterogeneity that provided indigenous representatives with access to decision makers – are unique to the Arctic Council, and indicate that other international institutions could not have triggered the same effect.

Consideration of the advantages that Arctic institutions may have over other levels of international governance is also relevant when evaluating the appropriateness of a catalytic approach to POPs regulation – that is, providing inputs to broader, ongoing processes rather than seeking to develop stronger regional rules. Major sources of air- and waterborne contaminants that bio-accumulate in the Arctic lie outside the region, so any regulatory initiative involving only Arctic states could not be adequate. Thus, whereas Arctic institutions have been better placed than broader ones to generate specific information about the threats to Arctic ecosystems, fauna and human health that are posed by hazardous substances, and thus muster support for action, the converse holds when it comes to the potential effectiveness of stronger international rules.

Also in other issue areas it would be unrealistic to expect Arctic institutions to engage in ambitious regulatory work. As Offerdal notes in Chapter 7, the strategic significance of oil and gas resources has meant reluctance among Arctic petroleum states to place binding regulation of such matters on the Arctic agenda. Another case in point is global warming. Although the 'Arctic eight', taken together, are responsible for more than half of the world's greenhouse gas emissions, this is essentially a global challenge, and moreover one that has been addressed under UN auspices for more than a decade. Findings generated under ACIA have been factored into the broader assessment work under the Intergovernmental Panel of Climate Change (IPPC), whose Fourth Assessment Report will include a separate chapter on the Arctic.

Although such cognitive contributions predominate, there is a normative element to the fact finding conducted under ACIA – it remains vague, however, and involves long and uncertain causal chains. On US insistence, the policy-recommendation part of the assessment was conducted at Senior Arctic Official (SAO) level rather than by scientists and science administrations, as originally envisaged. Yet, despite US scepticism, especially to those parts of the Policy Document that emphasized the gravity of the climate challenge and the need for domestic mitigation activities, the document was adopted and endorsed in the 2004 Ministerial Declaration. The ACIA reports and the Policy Document, the latter containing some of the clearest statements subscribed to by the present Bush administration on the need for action on global warming, have been widely disseminated in the USA, where policy makers and the general public have traditionally viewed 'climate science' with scepticism.[10] Distinctive features of the Arctic Council, especially the wide recognition of its environmental monitoring apparatus and its long-standing emphasis on indigenous concerns, combined to give saliency to these reports. The leading role played by US scientists in the assessment work, and the fact that indigenous peoples of Alaska are singled out among those most heavily and immediately affected by global warming, also explain why ACIA was received rather positively in this key climate country. The ACIA reports put a face on the climate problem in the USA and were subject to relatively greater media attention than the far more comprehensive assessment reports produced by the IPCC. That said, there is a long way to go from a positive reception of climate-impact reports to actual modification of US positions on international climate commitments.

In Chapter 5, we argued that Arctic Council activities soon outshone the sub-regional structures in Arctic environmental affairs. In the Council of Baltic Sea States, environmental collaboration has not had any strong Arctic dimension. For its part, the BEAR started out ambitiously with the establishment of an Environmental Task Force charged with developing multilateral environmental capacity-building initiatives in areas such as nuclear safety and reduction of industrial pollution in Murmansk and Arkhangelsk *oblasti* (Stokke 1994; Scrivener 1995). However, since state funding of proposed projects remained voluntary and was decided on a case-by-case basis, the lack of substantial financial contribution from countries other than Norway reduced the practical significance of decisions made at the sub-regional level. In practice, the centre of gravity for international capacity-enhancement decisions in the area returned to the bilateral level of governance or to fora that allocate funds from the EU or the USA. Under the Arctic Council, capacity enhancement was not salient in the early years. As we noted in Chapter 5, however, both the Regional Programme of Action for Protection of the Marine Environment from Land-Based Activities, which emphasizes development and implementation of Russia's National Plan of Action, and the Arctic Council Action Plan (ACAP) to Eliminate Pollution in the Arctic place considerable emphasis on the need for projects that can reduce regional discharges of harmful substances. ACAP in particular has produced tangible results, including the collection and safer storage of large amounts of PCBs and obsolete pesticides in Northwestern Russia, and the introduction of methods for cleaner production in the metallurgical complex in Norilsk.

Indigenous affairs

The perception that new tools were needed in order to alleviate socio-economic and other problems among its relatively large indigenous population was among the foremost motivations for Canada to launch the initiative for the Arctic Council. Many of the capacity-enhancement projects developed under its Sustainable Development Working Group address indigenous issues in particular. In Chapter 3, Wilson and Øverland argued that among the most important and lasting contributions of the Council has been to substantiate the special vulnerability of indigenous peoples to certain hazardous substances discharged mostly outside the region, and to raise awareness about those vulnerabilities in decision-making arenas that are broad enough to act effectively on them. This has been achieved primarily by ensuring that monitoring and research programmes, especially AMAP, include indicators and thematic foci that are relevant to indigenous concerns. In turn, indigenous organizations have used the widely accepted research findings emerging from these activities to develop their policy and arguments.

There is no doubt that, compared to other international bodies relevant to indigenous affairs, such as the ILO, the Arctic Council has been particularly well placed to assume such a fact-finding role. This is due mainly to the Council's broad functional scope, whereby studies of how indigenous peoples are affected by global environmental change have been able to benefit from a comprehensive environmental monitoring programme set up primarily for other purposes. More broadly, Wilson and Øverland argue, the Arctic institutions have created 'thematic bundles' in which indigenous matters are tied to issues higher up on the political agendas of Arctic states.

That the Arctic Council has emphasized knowledge generation does not mean that it has been irrelevant to normative advances in matters of indigenous concern – only that such advances have not been decided at the Arctic level. The clearest instance of Arctic institutions being used in efforts to catalyze normative change in other institutions in areas of indigenous interest are the POPs cases outlined above, in which the access of Inuit organizations to Canadian, and later US, foreign policy makers were important. According to Wilson and Øverland, that normative impact would not have been likely if the USA, whose large and complex foreign bureaucracy is less accessible to indigenous organizations, had occupied the driver's seat with respect to the Council's involvement in international POPs politics. More generally, by demonstrating that the physical health and ways of life of Arctic indigenous peoples are highly vulnerable to certain types of trans-boundary pollution, ice decline, and other impacts of global environmental change, Council activities have enabled indigenous organizations to forge a credible link between environmental protection issues and global norms on human rights.

That said, Wilson and Øverland point out that neither the Arctic Council nor the BEAR have paid much attention to the most immediate problem faced by indigenous peoples in northern Russia, Scandinavia and elsewhere: unrecognized or poorly implemented rights to land and water resources. Might such avoidance, as in the case of POPs regulation, be accounted for in terms of niche advantages based on the distinctive capacities of Arctic institutions? Although conceivable, this is by no means self-evident. True, indigenous organizations do pursue land-right issues under the UN Working Group on Indigenous Populations and there is some merit to the argument that this aspect of Arctic indigenous affairs can be appropriately addressed within the larger context of Fourth World politics. On the other hand, the participatory heterogeneity of Arctic institutions, with their blend of indigenous representatives and high-level government officials, could make these fora relevant for such matters as well. To a large extent, this blind spot of Arctic institutions regarding indigenous affairs reflects the hesitation of leading states to let such domestically contested issues, closely linked to extraction of oil, gas and mineral resources and not as clearly trans-boundary as pollution or threatening epidemics, be subject to political deliberations in intergovernmental bodies.[11]

Combating communicable diseases

In the area of public health care, most of the activity generated by Arctic regimes has occurred at the sub-regional level. In Chapter 4 Rowe and Hønneland showed how programmes under the Barents and Baltic Sea Regions, and recently also under the EU's Northern Dimension initiative, have strengthened the capacity in Northwest Russia and the Baltic states to cope with alarming rises in certain diseases that might spread across boundaries, especially HIV and tuberculosis. As to the Arctic Council, its work in the health sector has largely been confined to fact-finding activities under the AMAP.

The focus of sub-regional Arctic institutions on capacity enhancement exploits three competitive edges enjoyed by the Barents and Baltic Sea Regions over other international bodies that might play a role in connection with public health in Northwest Russia and the Baltic states. First, placing capacity enhancement initiatives at the Arctic (rather than the global) level pinpoints health differentials that are alarming enough to trigger action – whereas they would be dwarfed if placed alongside those that motivate costly WHO programmes in Third World countries. Second, nesting the health collaboration in regional initiatives with objectives that go well beyond health care – ultimately the involvement of post-Soviet states in European cooperative structures – has made possible Western funding for health purposes in Russia and the Baltic that would otherwise have been very difficult to obtain. From the perspective of region builders in the BEAR and the Council of Baltic Sea States, the near-collapse of the public health systems in those regions, associated with the transition from planned to market economies in the early 1990s, provided opportunities to demonstrate the ability of those new structures to produce tangible and visible benefits for Northern residents. Even with this additional motivation, it has not been easy to secure the financial basis for health programmes. As noted in Chapter 4, the Task Force on Communicable Disease Control in the Baltic Sea Region received considerably less for the entire five-year period of its existence than its initiators had envisaged for only the first year, and large regional states like Germany failed to contribute funds.

Finally, the participatory heterogeneity of Arctic institutions has provided direct access to province-level officials and health personnel, indicating that funds and proposed alterations in medical diagnostic and treatment procedures reached the ground level without being filtered through the central authorities. This feature is especially salient since resistance to behavioural change has been stronger at the central level than in the provinces, as we saw in the case of the combat against tuberculosis in Northwest Russia. Among the most significant changes brought about by the Baltic Sea Task Force and the Barents Health programme was the regional implementation of the WHO tuberculosis strategy, widely considered as more cost-efficient than the traditional Russian approach.[12] Cost efficiency is a major advantage because of severe cuts that had been implemented in health budgets in Northwest Russia and the Baltic states, but that strategy has met with considerable resistance from the central medical establishment in Russia. This was in part because the strategy, originally developed for Third World countries, had not been adequately adapted for application in a relatively advanced – albeit troubled – health structure. It was therefore dismissed by many as a 'Western' (i.e. non-Russian) approach inappropriately presented as a 'magic formula'. At the regional level, however, no similar opposition was voiced. Instead, the new diagnostic and treatment scheme championed under Arctic institutions was perceived as a useful new tool which – importantly and unlike many procedural instructions arriving from Moscow or St. Petersburg – was accompanied by the financial resources needed for implementation.

Rowe and Hønneland point out that, although the Barents Health Programme identified as priority areas such lifestyle-related health problems as alcoholism and smoking, most actual programme activities concerned communicable diseases. This might be explained in niche terms as capacity enhancement being limited to trans-boundary issues, where direct interdependences are involved. However, scanty attention to non-communicable health problems has characterized global assistance initiatives as well, including the Millennium Development Goals (Rechel *et al.* 2004), so also in those areas regional initiatives would have helped to fill a void.

Due to the causal complexities involved, it is difficult to determine how much of the recent stabilization and (in areas like tuberculosis occurrence) improvement in the state of public health in Northwest Russia and the Baltic states is due to programmes under Arctic institutions. Rowe and Hønneland show, however, that the Barents and Baltic Sea Region initiatives raised fresh funds and permitted coordination of programmes that induced more cost-efficient approaches to fighting communicable diseases in the area. Those new approaches originated in institutions beyond the region, but implementation would not have occurred without the programmatic contributions of the Arctic institutions.

Political mobilization

Among the striking features of Arctic institutions is their openness to actors that do not usually play prominent roles in international diplomacy – notably province-level governments, civil society groups and business organizations. Many have held high hopes that these institutions may empower new actors and contribute to a broadening of actual participation in decision making on Arctic affairs. This section sums up experiences thus far.

Indigenous peoples

Given the emphasis of the Arctic Council on indigenous representation and environmental monitoring, it is not surprising that it is the Arctic indigenous peoples and various scientist and expert communities that are most directly affected by this new institution. As argued by Wilson and Øverland in Chapter 3, this has been especially visible in the case of those indigenous groups whose organizations had not been oriented towards trans-national or international affairs. The formation of the Arctic Athabaskan Council and the Gwich'in Council International were direct consequences of the opportunities afforded by the Arctic Council for representation in an intergovernmental body whose deliberations and projects could be significant for these groups.

Even indigenous organizations with a strong tradition of trans-national work, notably the ICC and the Saami Council, have been further empowered by their status as Permanent Participants in the Council. This status involves practical access to the negotiations table on all issues addressed by the high-level forum. The ICC in particular has been receiving considerable attention from various governmental actors who perceive this organization as a potentially useful partner, or even ally, in international deliberations on Arctic affairs. Although participatory heterogeneity also marks the BEAR, indigenous representation here is limited to the province-level Regional Council. The latter body is salient for developing project work-priorities but does not permit, to the same extent as the Arctic Council, the development of contacts and partnerships with governmental decision makers.

An interesting feature of the Arctic Council is its institutional dynamism with respect to involving participants relevant to its activities. At first only three indigenous organizations were given status as Permanent Participants. The subsequent inclusion of the recently formed Athabaskan, Gwinch'in, and Aleut organizations indicates that the Arctic Council has not fallen prey to a malady often attributed to intergovernmental institutions: it has not preserved or enhanced initial differences among domestic groups in terms of influence by involving only the leading organizations. The conditions for such dynamism were favourable, however, not least since the case for openness to newcomers was consistently and forcefully championed by the USA.[13]

Beyond its formal inclusiveness of indigenous organizations, the Arctic Council has also triggered efforts to enable actual participation. The Indigenous Peoples' Secretariat, located in Copenhagen, was set up to assist indigenous representatives in their preparations for meetings and other activities under the Council (Bloom 1999: 719). Moreover, for firmly established indigenous organizations like the ICC and the Saami Council, the status of Permanent Participants has yielded incentives and opportunities to engage in capacity-enhancement work with the nascent indigenous movement in Russia.

Indigenous involvement has also been considerable in the environmental monitoring and impact assessment. The AMAP in particular has provided a vehicle for indigenous organizations to ensure that parameters of special importance to them have been included in data collection and research. In the ACIA, several Permanent Participants were involved in the Assessment Steering Committee, and programme work was tailored to take account of indigenous experience and knowledge on climate change. Several chapters in the Science Report deal with on indigenous issues in particular.

Participatory capacity enhancement is also within the mandate of the BEAR Working Group of Indigenous Peoples. Although the only working group to have been in continuous existence since the mid-1990s, its dynamism has been moderate and, as noted by Wilson and Øverland, the group itself has recently complained that the funding level is thoroughly inadequate to enable it to implement its work programme.

Scientists and experts

Environmental researchers are another actor category favourably affected by the niche orientation that characterizes Arctic institutions. As argued in Chapter 5, the collaborative monitoring activities coordinated under the Arctic Council, especially by AMAP but also by the working group on Conservation of Arctic Flora and Fauna (CAFF), have induced Arctic states to allocate more resources for such purposes than before. Arctic institutions have enabled the development of fairly stable trans-national networks, maintained through working- and expert-group participation and drawing on an impressive number of scientists, experts and research institutions. Each of the two comprehensive AMAP Assessment Reports involved more than three hundred researchers whose outputs have attracted considerable attention from bureaucratic and political decision makers as well as the general public. Similar numbers of climate researchers in various disciplines have contributed to the ACIA project. The network thereby created, as Hoel argues in Chapter 6, now amounts to a trans-national scientific constituency on Arctic climate issues with a remarkably high level of agreement on the nature and severity of the regional risks flowing from global warming. Through ACIA, international funds were also made available to involve a sizeable number of Russian experts, whose participation might otherwise have been constrained by the financial weakness of many research institutions in that country today.

Albeit on a smaller scale, the Oil and Gas Assessment has provided similar networking opportunities and is likely to enhance the societal visibility of research on hydrocarbon-environment relationships. As argued by Offerdal in Chapter 7, that study shows how deliberations under Arctic institutions can change state attitude to collaborative assessment projects. Although at first negative to the initiative, the USA ultimately assumed lead responsibility for two of the key science chapters, which will contribute to the prominence and visibility of the Oil and Gas Assessment. Similarly, data input from Russia had been low in the first part of the project, but reactions to this triggered stronger involvement of the Russian Ministry of Education and Science, and the situation improved markedly. In other words, the initially lukewarm responses of some important powers failed to slow down or stop an assessment that can be expected to draw considerable attention to Arctic oil and gas issues.

As pointed out by Rowe and Hønneland in Chapter 4, the involvement of regional health officials and medical expertise in broader networks is among the major and most lasting achievements of the health collaboration in the Barents and Baltic Sea Regions. Those collaborative programmes have strengthened lines of direct communication between regions and across health sectors within Russia. For the duration of those programmes, financial flows through the same networks have meant greater independence from central health authorities, especially in Northwest Russia. Whether such networks will be able to thrive after the sub-regional programmes have been completed, however, remains an open question.

Drivers and limitations

Although the Arctic institutions have been open to other societal organizations besides indigenous peoples and issue-specific experts, there is little evidence that they have significantly triggered greater participation in Arctic affairs by environmental groups and business representatives. In the Arctic Council, only the Advisory Committee on Protection of the Seas (ACOPS) and the World Wide Fund for Nature (WWF) has participated on a regular basis at meetings and contributed to programme activities.[14] One reason is that participation is costly and enabling funds have been directed at indigenous peoples' representatives. Moreover, deep and regular involvement in Council activities can be difficult to reconcile with the position of a critical outsider relative to state or province-level governments whenever there is disagreement about project contents.[15] A presumably even weightier explanation for the lack of a broad mobilizing effect of Arctic institutions on environmental and industry-group participation is their niche orientation toward non-controversial matters like fact finding and capacity enhancement, rather than politically contested issues related to regulation of economic or military activity. For instance, although the

WWF, through its newsletter *Arctic Bulletin*, disseminates information about Arctic Council and BEAR activities to environmental communities throughout the Arctic, Offerdal reports in Chapter 7 that neither this organization nor other leading environmental groups make much reference to the Arctic Council when addressing oil and gas issues currently of great topical interest in several Arctic states, in the broader media.

In part because the Arctic Council has been seen as largely focusing on indigenous issues and environmental fact finding, the business sector has been even less involved in the work of this institution than have environmental groups. In the sub-regional institutions, industry actors have played a greater role but not within the issue areas examined here. The ACIA implied some change in this respect, since that assessment was designed to address not only physical and biological systems but also social and economic ones. Through this programme, several industry organizations from sectors such as fisheries and shipping became directly involved in the work of the Arctic Council.

In examining the significance of Arctic institutions to this empowerment of certain societal groups and actors, external factors must be taken into consideration. Most obviously, Gorbachev's reshuffling of the Soviet state and its foreign-policy orientation – and the subsequent dissolution of the USSR – enabled cross-border contacts at a level and scope previously unthinkable. To a substantial extent, therefore, regional institutions and increased trans-national interaction are both premised on the same historical watershed. Moreover, the growing attention to Arctic natural resources, and the rising appreciation of this region's vulnerability to climate change, would probably have highlighted environmental and indigenous issues even if there had been no new circumpolar and sub-regional institutions.

In the case of indigenous peoples, perhaps the clearest evidence that their empowerment is partially caused by broader international processes is the gradual improvement of the quality of their representation in Arctic institutions, as pointed out by Wilson and Øverland in Chapter 3. The role assigned to indigenous representatives in the 1993 BEAR structure was considerably more prominent than it had been in the various Nordic institutions set up in earlier decades; and it is still more prominent in the Arctic Council. This development reflects the growing success of the Fourth World movement, most notably through the ILO and the activities of the UN Working Group on Indigenous Populations.

That said, distinctive features of the Arctic institutions, and the niches they have opted for in the various issue areas, have also shaped political mobilization in Arctic affairs. Although the tendency had been toward greater indigenous involvement, the participatory heterogeneity of the Arctic Council was no necessary outcome. Right from the outset, the US had been sceptical to the important role assigned to indigenous peoples in the Canadian proposal for an Arctic Council. That position was reasserted after the formation of the Council during the protracted negotiation of the rules on procedure (Archer and Scrivener 2000). This attitude was in part due to worries in the influential US environmental movement that indigenous rights might be used to counter the country's restrictive national legislation on marine-mammal protection (Keskitalo 2004: 72). The other leading Arctic state, Russia, had fewer direct reservations about profiled indigenous representation in the Council, but it was no driving force either: in effect, Russia supported US attempts to dilute the special status of indigenous representatives when elaborating the rules of procedure (Scrivener 1999). Hence, the participatory heterogeneity of the Arctic Council is the outcome of a process that could have yielded a different structure. It was Canada, a middle-sized power in Arctic affairs, and to some extent also Denmark, that pressed for and ultimately managed to obtain a strong indigenous stamp on the new institution.

The financial basis for enabling participation at meetings or in programmes, and for costly environmental monitoring and public health programmes, has been secured by the ability of Arctic institutions to trigger funds from the wealthier Arctic states to support activities whose main impacts will be felt in other parts of the region. The relatively high agenda ranking of indigenous affairs in Canada and Denmark goes far in explaining why the Indigenous Peoples' Secretariat has been funded largely by those two states (Langlais 2000: 14). Similarly, we saw above that monitoring and capacity enhancement in the environmental and health sectors has been fuelled by the broader objectives, especially among Nordic states, on creating sustainable political infrastructure and economic integration across the former East–West boundary.

Arctic region building

Regularized meetings among decision makers at political and bureaucratic levels, in combination with substantial ground-level, transnational networks on indigenous, health, science, and environmental affairs, are important steps in the building of a political region. As argued in Chapter 2, whether the Arctic qualifies as a 'political region' is more than an academic question. Regionality is in part a matter of functional interaction, but it also concerns whether regional actors see the challenges they face as distinctively Arctic and warranting regional responses – rather than geographically narrower or broader ones. The question of regionality thus impinges on the political and socio-economic bases for pursuing collaboration among Arctic states, provinces and civil society organizations.

There is much to indicate that the high prominence of indigenous affairs in the Arctic Council has enhanced the perception in broader fora that the Arctic is a distinctive region in need of special attention and separate treatment. This highlights the discursive side of regionality – the extent to which the Arctic is seen and spoken of as unit. As pointed out by Wilson and Øverland in Chapter 3, especially the ICC has promoted the idea in various international organizations and conferences, including in the UN, that the Arctic is tied together by the presence and particular vulnerability of indigenous peoples to global environmental change. As noted above, the link between environmental protection and indigenous human rights is made explicit in such contexts and with increasing persuasiveness.

Keen awareness of the discursive aspect of regionality is evident in the many BEAR projects that involve 'regional statements' – visible institutional expression of regional community. Consider for instance Norway's establishment of the Barents Secretariat in Kirkenes, the city that hosted the founding meeting in 1993; and more recently the creation of a Barents Institute to conduct multidisciplinary research on regional affairs. Similarly, some of the most costly and highly profiled BEAR projects on indigenous issues have targeted buildings and activities at the Choom National Cultural Centre in the Russian Saami 'capital' of Lovozero – which might reflect an emphasis on activities with high *symbolic* content. Indeed, in Chapter 3 Wilson and Øverland question the instrumental, problem-solving contributions of these and similar projects. In Chapter 5 we also note that in Barents Region rhetoric, any distinction between accomplishments under bilateral or trilateral collaborative vehicles and those arising from BEAR bodies has tended to be under-communicated. This may be seen as an effort to maximize the region-building effects of any tangible results of regional collaboration, whether bilateral or multilateral.

Even more striking is the impact on discursive regionality that arises from singling out the Arctic as a key region for monitoring and research in such salient issue areas as POPs and climate change. The ACIA's Overview Report and Policy Document disseminated and lent credibility to information about climate change that will affect monitoring and research activities among Arctic states; it might even affect the priority given to mitigation and adaptation activities in some countries.

As Hoel pointed out in Chapter 6, those reports also gave the Arctic Council more media attention than any previous event, in part due to the tangibility and visibility of climate impacts in this region. The 2004 Arctic Council Ministerial was the first to be attended and extensively covered by global news agencies, and ACIA outputs continue to provide substance for widely broadcast news and feature articles on the relationship between the Arctic region and global warming. In contrast, as argued by Offerdal in Chapter 7, the low profile of Arctic institutions so far with respect to oil and gas development has precluded any strong impacts on region building. That said, the growing political saliency of Arctic petroleum resources is now encouraging regional states to pay greater attention to their northerly territories, and activities such as the Oil and Gas Assessment should contribute to a framing of issues surrounding those resources as distinctly 'Arctic'.

As to the Baltic health collaboration, Rowe and Hønneland maintain in Chapter 4 that its placement in a multilateral, sub-regional framework proved highly useful by enabling closer ties between the medical communities in Russia and in the Baltic states. Those ties had been severely strained by the general animosity that followed the short-lived Soviet occupation of the Baltic capitals in 1991, and there is much to suggest that the notable improvement in Russian–Baltic relations in the health sector since then can be credited to the external impulse provided by the Arctic sub-regional institutions.

Promising as these developments are, the Arctic cannot be said to constitute a political region in the strict meaning of the term. The sense of common identity among the four million residents of the area remains generally weak, as is public awareness of activities under Arctic institutions. For many problems, other governance levels will be more important than the Arctic one – whether because bilateral or sub-regional interdependencies are more intense, or because problem solving requires participation by a broader set of states. From a functional point of view, the Arctic is often either too big or too small. Despite this, there is no doubt that the institutions examined in this book have succeeded in establishing Arctic affairs as a distinctive international policy area that merits regular attention by decision makers at state, provincial and societal levels. Increasingly also, that distinctiveness is gaining recognition in broader international fora.

Conclusions

This chapter has summed up findings from five case studies of Arctic institutions and their impacts on regional connectedness, political involvement, and specific problem solving in issue areas that rank high on Arctic political agendas. Starting from a low level, functional and discursive regionality is now on the rise in the Arctic. The institutions examined here have contributed to the development and maintenance of networks that nurture both aspects. Interaction within such networks is broadened by the involvement of province-level authorities and civil society groups, including indigenous organizations. Discursively, the emphasis of the Arctic Council on circumpolar environmental monitoring and indigenous issues has directed greater attention – within the region, and beyond – to the Arctic dimension of some global issues, like hazardous substances and climate change. However, although circumpolar and the sub-regional institutions provide means for addressing common or similar challenges, other levels of governance will continue to offer equally or more powerful instruments on many issues.

Arctic institutions are the most effective – make the biggest difference – when they focus on activities or problem aspects where they enjoy niche advantages: where distinctive features of Arctic institutions make them better placed than others to extract or utilize the resources needed for problem solving. The cognitive, or fact-finding, niche is the one most widely chosen in the issue areas examined here, especially within the Arctic Council. This niche orientation makes it possible to reap the gains that arise from coordination of data collection and analysis among states governing adjacent territories with certain shared biophysical characteristics. It also exploits the preparedness of states otherwise opposed to the development of strong international institutions in the Arctic to at least support environmental monitoring.

With respect to normative contributions, the Arctic approach has been far more limited, largely echoing broader international regimes already in existence. In no cases have institutions examined here provided legally binding rules, or review procedures that could give political teeth to non-binding ones. Sometimes, but not always, such nesting is justified by the fact that broader regimes involve a more relevant set of actors, or already apply to all Arctic states. In the regulation of hazardous pollutants, Arctic institutions have served as platforms for efforts to influence spatially broader regulatory processes – partly by feeding in research findings on Arctic vulnerabilities, and partly by prodding Arctic states to take a more common stand on issues of concern. In the regulation of POPs and, to a lesser extent, in international climate politics, such catalytic efforts have made use of the environmental monitoring capacity of the Arctic Council and its external credibility as a acknowledged conveyor of indigenous concerns.

Finally, a capacity enhancement niche has been carved out in certain areas such as communicable diseases, cleaner production in process industries, and safer storage and treatment of hazardous waste. These are areas with a trans-boundary component, so that support from other Arctic states to projects in Russia and the Baltics has in part been driven by self-interest. However, the willingness to pay for such capacity-enhancement efforts also derives from the fact that Arctic institutions have been significantly fuelled by broader security and economic objectives, centred on the goal of involving – enmeshing – Russia and the Western Arctic states within common cooperative structures.

Notes

1 I would like to thank David Scrivener, Oran Young and my fellow contributors to this book for very helpful comments. By 'Arctic' international institutions is meant those that involve several Arctic states and have among their priority areas the challenges and opportunities that arise or become visible in the Arctic. We apply the spatial boundary of the Arctic defined in the *Arctic Human Development Report* (ADHR 2004: 17–18); see Chapter 1.

2 As noted in Chapter 1, this book does not aspire to cover the full range of priority areas defined under the relatively new Arctic institutions. The development of stronger economic ties among the northerly territories of the states involved, an important goal of especially the sub-regional institutions, is not addressed directly.

3 See the discussion of this in Scrivener (1999).

4 The comprehensive reports were published in 1997 and 2002, the health report in 2002; the oil and gas assessment is due in autumn 2006.

5 On nesting, see Aggarwal (1983) and Young (1996).

6 On the Guidelines for Transfers of Refined Oil and Oil Products in Arctic Waters, the Field Guide for Oil Spills Response, the International Murre Conservation Strategy and Action Plan and the Circumpolar Eider Conservation Strategy and Action Plan, see Chapter 5 by Stokke, Hønneland and Schei.

7 See Chapter 5; also Stokke (2000).

8 See Chapter 5.

9 For accounts and assessments of these cases of institutional interplay, see Downie and Fenge (2003).

10 See Chapter 6 by Hoel.

11 Depending on the level of specificity of any measure considered, indigenous organizations not currently active in the Arctic Council might also object to such issues being placed on the Arctic agenda; see below on the requirements for eligibility to be considered for status as Permanent Participant.

12 As outlined by Rowe and Hønneland in Chapter 4, the traditional Russian approach to tuberculosis detection and treatment differs from WHO's Directly Observed Treatment with Short-course Therapy (DOTS) strategy by

pursuing active case finding through mass screening of the population (instead of self-reporting patients); hospitalization and isolation (instead of out-patient treatment); long-term, individualized multi-drug approaches (instead of short-course standard cure); and the use of surgery no longer employed in the West.

13   Only organizations that represent an Arctic indigenous people resident in more than one Arctic state or more than one Arctic indigenous people resident within a single Arctic state are eligible for consideration as Permanent Participants in the Arctic Council; see Declaration on the Establishment of the Arctic Council, Art. 2, and Rules of Procedure, Part V.

14   See Chapters 5–7; these organizations also have formal observer status within the Arctic Council. On WWF contributions to work under CAFF and the working group on Protection of the Marine Environment (PAME), see Archer and Scrivener (2000).

15   See Chapter 7 by Offerdal. For a discussion of this dilemma in the context of global climate politics, see Andresen and Gulbrandsen (2005).

References

AHDR (2004) *Arctic Human Development Report*. Akureyri: Stefansson Arctic Institute, under the auspices of the Icelandic Chairmanship of the Arctic Council, 2002–2004.

Aggarwal, Vinod K. (1983) 'The unraveling of the Multi-Fiber Arrangement, 1981: an examination of international regime change', *International Organization*, 37: 617–645.

AMAP (Arctic Monitoring and Assessment Programme) (1997) *Arctic Pollution Issues: A State of the Arctic Environment Report*, Oslo: Arctic Monitoring and Assessment Programme. www.amap.no.

AMAP (Arctic Monitoring and Assessment Programme) (2002) *Arctic Pollution 2002*, Oslo: Arctic Monitoring and Assessment Programme. www.amap.no.

AMAP (Arctic Monitoring and Assessment Programme) (2004) *AMAP Strategy 2004+*, Oslo: Arctic Monitoring and Assessment Programme. www.amap.no.

Andresen, Steinar and Lars H. Gulbrandsen (2005) 'The role of green NGOs in promoting climate compliance' in O. S. Stokke, J. Hovi and G. Ulfstein (eds.), *Implementing the Climate Regime: International Compliance*, London: Earthscan , pp. 169–186.

Archer, Clive and David Scrivener (2000) 'International cooperation in the Arctic environment' in M. Muttall and T. V. Callaghan (eds.), *The Arctic: Environment, People, Policy*, Amsterdam: Harwood Academic, pp. 601–620.

Arctic Council, Rules of Procedure, As adopted by the Arctic Council at the First Arctic Council Ministerial Meeting, held in Iqaluit, Canada, 17–18 September 1998. www.arctic-council.org.

Bloom, Evan T. (1999) 'Current Developments – Establishment of the Arctic Council', *American Journal of International Law*, 93: 712–722.

Declaration on the Establishment of the Arctic Council (Ottawa, 19 September 1996). www.arctic-council.org.

Downie, David L., and Terry Fenge (eds.) (2003) *Northern Lights Against POPs: Combatting Toxic Threats in the Arctic*, Montreal: McGill–Queen's University Press.

Fenge, Terry (2003) 'POPs and Inuit: influencing the global agenda' in D. L. Downie and T. Fenge (eds.), *Northern Lights Against POPs: Combatting Toxic Threats in the Arctic*, Montreal: McGill–Queen's University Press, pp. 192–213.

Huntington, Henry P. and Michelle Sparck (2003) 'POPs in Alaska: engaging the USA' in D. L. Downie and T. Fenge (eds.), *Northern Lights Against POPs: Combatting Toxic Threats in the Arctic*, Montreal: McGill–Queen's University Press, pp. 214–223.

Keskitalo, E. C. H. (2004) *Negotiating the Arctic: The Construction of an International Region*, New York: Routledge.

Langlais, Richard (2000) *Arctic Co-operation Organisations: A Status Report*, Standing Committee of Parliamentarians of the Arctic Region. www.arcticparl.org.

PAME (Working Group on Protection of the Marine Environment) (1996) *Report to the Third Ministerial Conference on the Protection of the Arctic Environment, Inuvik, Canada, 20–21 March 1996*. www.arctic.council.org.

Rechel, B., L. Shapo and M. McKee (2004) *Millennium Development Goals for Health in Europe and Central Asia: Relevance and Policy Implications*, Washington, DC: World Bank Working Paper No. 33.

Scrivener, David (1995) 'Environmental cooperation in the Euro-Arctic', *Environmental Politics*, 4: 320–327.

Scrivener, David (1999) 'Arctic environmental cooperation in transition', *Polar Record*, 35: 51–58.

Stenlund, Peter (2002) 'Lessons in regional cooperation from the Arctic', *Ocean and Coastal Management*, 45: 835–839.

Stokke, Olav Schram (1990) 'The Northern environment: is cooperation coming?' *Annals of the American Academy for Political and Social Science*, 512: 58–69.

Stokke, Olav Schram (1994) 'Environmental Cooperation as a Driving Force in the Barents Region' in O. S. Stokke and O. Tunander (eds.), *The Barents Region: Cooperation in Arctic Europe*, London: SAGE, pp. 145–159.

Stokke, Olav Schram (2000) 'Sub-regional cooperation and protection of the Arctic marine environment: the Barents Sea' in D. Vidas (ed.), *Protecting the Polar Marine Environment: Law and Policy for Pollution Prevention*, Cambridge University Press, pp. 124–48.

Young, Oran R. (1996) 'Institutional linkages in international society: polar perspectives', *Global Governance*, 2: 1–24.

Index

eBooks – at www.eBookstore.tandf.co.uk

A library at your fingertips!

eBooks are electronic versions of printed books. You can store them on your PC/laptop or browse them online.

They have advantages for anyone needing rapid access to a wide variety of published, copyright information.

eBooks can help your research by enabling you to bookmark chapters, annotate text and use instant searches to find specific words or phrases. Several eBook files would fit on even a small laptop or PDA.

NEW: Save money by eSubscribing: cheap, online access to any eBook for as long as you need it.

Annual subscription packages

We now offer special low-cost bulk subscriptions to packages of eBooks in certain subject areas. These are available to libraries or to individuals.

For more information please contact webmaster.ebooks@tandf.co.uk

We're continually developing the eBook concept, so keep up to date by visiting the website.

www.eBookstore.tandf.co.uk